Laércio Pereira de Oliveira

Carbon and nutrients in the Mixed Ombrophylous Forest and Pinus Elliottii

Laércio Pereira de Oliveira

Carbon and nutrients in the Mixed Ombrophylous Forest and Pinus Elliottii

Contributions to sustainable forest management

ScienciaScripts

Imprint
Any brand names and product names mentioned in this book are subject to trademark, brand or patent protection and are trademarks or registered trademarks of their respective holders. The use of brand names, product names, common names, trade names, product descriptions etc. even without a particular marking in this work is in no way to be construed to mean that such names may be regarded as unrestricted in respect of trademark and brand protection legislation and could thus be used by anyone.

Cover image: www.ingimage.com

This book is a translation from the original published under ISBN 978-3-330-99699-1.

Publisher:
Sciencia Scripts
is a trademark of
Dodo Books Indian Ocean Ltd. and OmniScriptum S.R.L publishing group

120 High Road, East Finchley, London, N2 9ED, United Kingdom
Str. Armeneasca 28/1, office 1, Chisinau MD-2012, Republic of Moldova, Europe
Managing Directors: Ieva Konstantinova, Victoria Ursu
info@omniscriptum.com

Printed at: see last page
ISBN: 978-620-8-60785-2

To God, my rock and my shield, my support in times of anguish, my presence in times of joy.

To Jesus Christ, the author of my faith.

To my parents Tércio Lemenhe de Oliveira (in memoriam) and Léa Pereira de Oliveira (in memoriam), who gave up many dreams in my favour.

To my wife Fátima and my children Enrico and Chamille, for their encouragement and affection, for nurturing my capacity to love.

ACKNOWLEDGEMENTS

My thanks go to Professor Kátia Cylene Lombardi, my supervisor, for her patience and support during my work.

To Professor Marcos Vinicius Winckler Caldeira, for his significant contribution.

To Professors Andrea Nogueira Dias, Mário Takao Inoue and Luciano Farinha Watzlawick, for their support in the development of this work.

To Professor Mário Humberto Menon, Director of the Unicentro-Irati Campus.

To Prof[1]. Sandra Regina de Oliveira Garcia, Head of the Department of Education and Labour/Secretary of State for Education - PR, for her words of encouragement.

To the professors of the master's programme in forestry sciences at Unicentro, for another stage in my professional career.

To my colleague and friend Eng. Wanderley Carlos Perdoncini, for his support and companionship.

To researcher Kauana Dickow (UFPR), for her technical help.

To the Department of Architecture and Urbanism/Geoprocessing of Irati City Hall for their cartographic assistance, especially Sérgio Caetano.

To my friend Vivian Dallagnol de Campos, for her help in structuring the work.

To my fellow master's students for their companionship, especially Alex Sawczuk for his help and partnership in the work.

To the students of Colégio Florestal, Wesley Batista Amann, Luiz Fernando Cordeiro and Romeu Pontes Junior, for their collaboration.

SUMMARY

The soil-plant relationship is very important for understanding the development of planted forests and natural forests, considering that trees interfere with the soil through their root system and the deposition of organic material. The soil, in turn, is directly responsible for supplying plants with nutrients, retaining water, providing shelter for fauna, as well as being the natural substrate for various plant formations. The aim of this study was therefore to analyse some of the nutrient and organic carbon cycling in the Montane Mixed Ombrophilous Forest and in a 33-year-old Pinus elliottii Engelm plantation. The study area is located in the President Costa e Silva State Forestry College, covering 176.5 ha. The representative soils of both forests were classified as Dystrophic LITHOLIC NEOSSOLO (Pinus elliottii Engelm) and Typical Alumic HUMIC CAMBISSOLO (FOMM). The results obtained from this research show that the carbon and nutrient concentrations of the soil and litter vary significantly in the two forest formations: planted (Pinus elliottii Engelm) and native (Montane Mixed Ombrophilous Forest). In the forest plantation, the carbon content, as well as the macro and micronutrient content, showed average values below those found in the native forest. The diversity of species and the greater ecological complexity of the Montane Ombrophilous Mixed Forest may have influenced the results found, compared to the Pinus elliottii Engelm plantation. It was concluded that the Montane Mixed Ombrophilous Forest (MCOMF) presented better conditions for supplying essential nutrients and fixing organic carbon.

KEYWORDS: Planted Forests, Natural Forests, Nutrients, Organic Carbon.

SUMMARY

CHAPTER 1	5
CHAPTER 2	6
CHAPTER 3	7
CHAPTER 4	22
CHAPTER 5	31
CHAPTER 6	42
CHAPTER 7	43

CHAPTER 1

INTRODUCTION

Brazilian forestry has a lot of information on forest plantations, especially of the Pinus and Eucalyptus genera, but there is a lack of research on native forest species, in experimental open-canopy or intercropped plantations. Forest management of native species is still in its infancy and does not provide many answers regarding the use of wood and non-timber forest products, or the most suitable sites for a given species.

Paraná's native tree species are increasingly restricted to smaller areas due to various factors, such as the expansion of agriculture and livestock farming, the growth of urban areas and indiscriminate deforestation. As a result, there is a progressive loss of biodiversity, with concrete possibilities of the extinction of some species. This is why it is essential to be concerned about the wealth contained in native forests, which are also responsible for regulating the water cycle and preserving important species of wild fauna.

In this context, it is essential to study and identify the levels of mineral nutrients, which play a fundamental role in tree growth, and whose dynamics of adsorption in the soil and absorption by plants are influenced by the volume and quality of the litter, as well as the decomposition conditions of this material.

Currently, due to climate change as a result of the greenhouse effect, a lot of work is focused on the issue of atmospheric carbon sequestration. Thus, forestry is gaining importance, both in the preservation of native forests and in the establishment of forests for productive purposes, as well as the recovery of degraded areas.

In view of these observations, the aim of this research was to evaluate carbon sequestration in the biomass of the leaf litter and in the soil, comparing sites of Montane Mixed Ombrophilous Forest and Pinus elliotti Engelm plantations under different soil classes (Cambissolo and Neossolo), both with poorly developed profiles, since in soils the dynamics of organic matter, and consequently carbon, is related to vegetation cover, climatic, microbial and soil conditions.

The management of native forests, forest plantations and planting experiments in the open or in shaded strips contribute to knowledge of the productive potential of forest species, since this potential is directly related to biomass production and carbon sequestration.

Thus, this study correlates: 1) the availability of nutrients for the development of forest species with the production of biomass in the accumulated litter, and 2) the formation of organic matter with soil properties in forest systems, with the aim of evaluating the sequestration of carbon in the soil and contributing to possible new directions in soil management practices.

CHAPTER 2

OBJECTIVES

2.1 GENERAL OBJECTIVE

1. To study part of the cycling of nutrients and organic carbon in the Montane Mixed Ombrophilous Forest and in a 33-year-old Pinus elliottii Engelm plantation.

2.2 SPECIFIC OBJECTIVES

1. Quantifying the accumulated litter in the Montane Mixed Ombrophilous Forest and in a Pinus elliottii Engelm plantation;
2. Quantify the nutrient and organic carbon content of the accumulated litter under the Montane Mixed Ombrophilous Forest and in a Pinus elliottii Engelm plantation;
3. To analyse the chemical and physical properties of the soil under the Montane Mixed Ombrophilous Forest and in a Pinus elliottii Engelm plantation.

CHAPTER 3

THEORETICAL FRAMEWORK

3.1 PINUS ELLIOTTII VAR. ELLIOTTII ("SLASH-PINE")

This species is one of the most important pines in the southeastern United States. Its dispersal zone extends from southern South Carolina (33.5° N latitude) to central Florida and southeastern Louisiana (30° N latitude). In these regions it grows on sandy soils, at altitudes of less than 990 m. It is characterised by a warm climate, with humid summers and lower rainfall in spring. The average annual rainfall is 1270 mm and the average annual temperature is 17.2°C, with occasional extremes of 41°C and -18°C (KRONKA et al., 2005).

In its natural range, it is restricted to altitudes of 0 to 150 metres. In tropical plantation areas, it is mainly planted at altitudes of 500 to 2500 metres. Pinus elliottii Engelm reaches heights of 20 to 30 m (maximum 40 m) and a diameter at breast height (DBH) of 60 to 90 cm. The slender trunk is very straight in most cases. The crown is clearly ovoid. The roots can penetrate the soil to a depth of 5 metres or more. The bark of young trees is greyish and deeply furrowed (LAMPRECHT, 1990).

According to Lamprecht, 1990, the shiny, dark green acicles reach a length of 17 to 25 cm, arranged in bundles of 2 or 3. They have 2 to 10 resin canals. The cones are shiny and greyish-brown in colour, 7 to 15 cm long, which - when they have closed scales - is at least twice as long as they are wide. The seeds are winged and are 6 x 3 mm, dark in colour and oval in shape.

As far as soils are concerned, Pinus elliottii var. elliottii has a natural preference for acidic, sandy soils, mainly located in lowlands and near watercourses, generally in areas where the water table is close to the surface (HARLOW et al., 1979). The slash-pine (the common name for Pinus elliottii Engelm in its region of origin) is very resistant to frost and largely tolerant of winds with high salt levels.

As a fast-growing heliophytic species, Pinus elliottii Engelm is highly competitive with grasses and woody shrubs. In natural forests, regular seed production begins from the age of 20, with 27,000 to 34,000 seeds Kg^{-1}. According to Webb et al. (1980), in addition to the United States, the largest seed exporters are South Africa and Australia.

In Brazil, it finds ideal growing conditions from Rio Grande do Sul to the centre of Paraná and the south of São Paulo. It can also be grown in higher altitude areas (Serra da Mantiqueira, do Mar, Bocaina and dos Órgãos). It requires evenly distributed rainfall throughout the year, cold winters and no water deficit (KRONKA et al., 2005).

3.2 PINE PLANTATIONS IN BRAZIL

The Pinus genus has gained an important place in Brazilian forestry, as it is the raw material for various products, such as pulp and paper, packaging, panelling and plywood, construction material, furniture and resin, with numerous applications in the chemical industries. The first studies on subtropical pines date back to 1936 by the São Paulo Forestry Institute, when the first seeds of Pinus elliottii var. elliottii and Pinus taeda L were introduced (KRONKA et al., 2005).

The formation of forest stands that meet the required quality standards begins with the production of selected seedlings, which depend on seeds of suitable genetic material that is compatible with local conditions. It is also important to emphasise that combating leaf-cutting ants is a mandatory operation, especially during planting or in the first years of growth. The wood wasp is currently a worrying pest that requires preventative and control measures (KRONKA et al., 2005).

Freitas (2005) apud Kronka et al., 2005) mentions that there were 1.8 million hectares of Pinus planted in Brazil up to that date. In Paraná, the planted area was equivalent to 605,130 ha (SBS, 2001).

Sustainable forestry production can be compared to planting on an annual and continuous basis, in order to adequately meet the demand for raw materials. In this way, economic results can be obtained over a long period of time, a characteristic prerequisite for investment in the activity. Therefore, the best forestry productivity results from using suitable genetic material that is compatible with local conditions. Soil preparation depends on its previous vegetation cover, whether or not there are roads, plot boundaries or firebreaks (KRONKA et al., 2005).

Also according to Kronka et al. (2005), the control of leaf-cutting ants is a mandatory operation, even in places where this pest does not seem to be very intense. Another important pest is the wood wasp (Sirex noctilo), which prefers to attack weakened trees, so proper management is the best form of control.

Spacing is one of the most important decisions in forestry implementation and there is no general rule, but the first spacings adopted at the time of the Tax Incentives were 1.5 x 1.5 m; 2.0 x 2.0 m and 2.0 x 2.5 m. Later, wider spacings were adopted: 2.5 x 2.5 m and 3.0 x 3.0 m (KRONKA et al., 2005).

The silvicultural practices of thinning and pruning are also widely adopted for plantations of Pinus elliottii Engelm. Thinning aims to remove the worst trees, giving good growing conditions to the remaining ones, which means special focus on initial spacing, one of the purposes of which is to establish sufficient conditions for successive selections. Pruning consists of eliminating branches up to a certain height, with the aim of promoting the formation of knot-free wood (KRONKA et al., 2005).

Stape (2003) indicates an Average Annual Increase (AIA) of 18 - 45 m^3 ha year $^{-1}$ for *Pinus elliottii* Engelm. In Irati - PR, the AMI was 32.0 - 39.3 m^3 ha year $^{-1}$ for *Pinus elliottii* var. *elliottii*, for some provenances.

The most traditional silvicultural systems for *Pinus* plantations are clear felling, successive felling and selective felling, using the high stem method (STAPE, 1996).

3.3 MIXED OMBROPHILOUS FOREST

The name Ombrophilous Forest was proposed by Muller Dombois and Ellemberg (1955-1956) to replace Tropical Rain Forest, suggested by Richards (1952), quoted by Lacerda (1999). However, both have the same meaning, i.e. "rain-friendly". The term Ombrophilous has Greek origins, while the term Pluvial has Latin origins and characterises tropical ecological physiognomies. Within the natural occurrence of the Ombrophilous Forest there are specific ecosystems, according to the soil and climate characteristics of each region.

The concept of Mixed Ombrophilous Forest comes from the occurrence of a mixture of different species, defining typical physiognomic patterns, in a climatic zone that is characteristically pluvial (IBGE, 1992). This type of forest is also known as Araucaria Forest, Pine Forest, Araucaria Forest and Subtropical Aciculifolia Forest (BRITEZ *et al.,* 1995). These names are due to the presence of *Araucaria angustifolia* (Benth.) O. Kuntze, which dominates the canopy and physiognomically characterises this forest formation. However, the term mixed is related to the mixture of floras (IBGE, 1991).

In Brazil, the Araucaria Forest, according to Klein (1960) and Hueck (1972) occurs in the states of Paraná, Santa Catarina, Rio Grande do Sul and southern São Paulo and, in isolated patches, in the states of Rio de Janeiro, Minas Gerais and Espírito Santo, also reaching the province of Missiones in Argentina and western Paraguay.

The Mixed Ombrophilous Forest has, according to Veloso et al. (1991), there are four distinct subformations: a) Alluvial: on ancient fluvial soils - *Araucaria angustifolia* associated with *Podocarpus lambertii* and *Drimys brasiliensis* or genera from the Lauraceae family; b) Submontane: from 50 to more or less 400 metres above sea level; c) Montane: from 400 to more or less 1000m altitude - *Araucaria angustifolia* associated with *Ocotea porosa* formed very characteristic groupings and d) High Montana: situated at more than 1000m altitude - *Araucaria angustifolia* associated with *Podocarpus lambertii, Drimys brasiliensis, Cedrela fissilis* and genera from the Lauraceae and Myrtaceae families.

The area of natural occurrence of the Mixed Ombrophilous Forest (FOM) extends in the Southern Region of Brazil, between latitudes 24° and 30° S, at an altitude ranging from 500 m to

1400 m above sea level (a.n.m.). It also occurs in disjunct areas in the Southeast of Brazil, between latitudes 18° and 24° S, at an altitude ranging from 1400 m to 1800 m a.n.m. At the highest altitudes in the South, this formation forms part of a vegetation mosaic made up of forests and natural grasslands (KLEIN, 1960). In this type of forest, *Araucaria angustifolia* (Benth.) O. Kuntze forms the upper canopy and is dominant in the vegetation, accounting for a large percentage of the individuals in the upper layer (LONGHI, 1980; LEITE and KLEIN, 1990).

Valério *et al.* (2008) point out that *Araucaria angustifolia* (Benth.) O. Kuntze is the main arboreal component of its stratum in the Mixed Ombrophilous Forest, accompanied by Bracatinga (*Mimosa scabrella* Benth.) and Yerba mate (*Ilex paraguariensis* A. St.-Hil.). St.-Hil.), currently exploited species with significant economic value in their distribution areas, with scattered remnants in the states of Rio Grande do Sul, Santa Catarina, Paraná and São Paulo.

The interior of the Mixed Ombrophilous Forest is made up of a fairly homogeneous stratum with the constant presence of *Araucaria* and *Podocarpus*. However, in their area of distribution, these two species have typical behaviour. *Araucaria angustifolia* is the species of the dry slopes and plateaus, although it also extends into valley soils with a shallow water table, where, however, it is less common than *Podocarpus lambertii*, the species of the valleys in humid soils (HUECK, 1972).

In addition to *Araucaria angustifolia* and *Podocarpus lambertii*, other tree species occur constantly in the association. According to Longhi (1980), these species come from high-altitude forests, which grow preferentially on the eastern slopes, rich in rainfall, including Ocotea porosa, Cedrela fissilis, Ilex paraguariensis, Balfourodendron riedelianum, Cabralea canjerana, Holocalyx balansae and others, such as Podocarpus sellowii, Ocotea pretiosa, Ocotea puberula, Ocotea catarinensis, Campomanesia xanthocarpa, Vernonia discolor.

At altitudes below 500 metres, the occurrence of Mixed Ombrophilous Forest is only observed on the slopes of valleys and erosion canyons, associated with Syagrus romanzoffiana (jerivá), in the cold run-off lines (HUECK, 1972).

In Paraná, the most common subformations of the Mixed Ombrophilous Forest are Montana and Altomontana, usually presenting the appearance of a pure association, due to the physiognomic dominance imposed by the araucaria (IBGE, 1990).

The study by Castela (2001) provides detailed information on forest cover with Araucaria in the state of Paraná. The extent of the Araucaria Forest was 8,295,750 ha, 41.5 % and 16.5 % respectively of the state's total area (19,972,926 ha). However, this quantification goes a little further than that reported by Maack (1968), when there were 7,378.00 ha of Araucaria Forest and 3,053.20 ha of grasslands.

According to Castela (2001), the south-central region of the state of Paraná has the greatest coverage of Araucaria Forest, where the municipalities of Bituruna, General Carneiro, Coronel

Domingos Soares, Porto Vitória, União da Vitória, Cruz Machado, Inácio Martins, part of Pinhão, Guarapuava and Turvo are located, the latter following the Serra da Esperança.

Another region with very significant forest cover is that which follows the Devonian Escarpment on the first plateau, where early-stage formations predominate. This region shows the greatest differences in terms of forest typology, probably due to the different past and current anthropogenic uses (CASTELA, 2001).

In addition, the Araucaria Forest region (CASTELA, 2001) is characterised by a temperate climate, where frost probably plays a fundamental and selective role in the occurrence of certain species, also influencing their physiology, such as leaf fall.

Figueiredo Filho et al. (2003), studying the seasonal increase in diameter of 7 species in a Mixed Ombrophilous Forest located in São João do Triunfo, in the southern region of the state of Paraná, in 131 trees, found, after three years of observations, that the highest rate of increase in diameter occurred in summer, followed by spring, autumn and winter, accounting for 50, 31, 12 and 7% of annual production, respectively. The species Prunus brasiliensis, Cinnamomum visiculosum and Nectandra grandiflora showed the highest rates of increase. The average annual increase in diameter of the 6 hardwoods studied was 0.261 cm; Araucaria with 0.129 cm; for the 7 species, this increase was 0.199 cm. The increase in diameter seems to be more strongly correlated with rainfall than with temperature. The 7 species were selected according to the criteria of abundance, dominance and commercial importance. A total of 62 Araucaria angustifolia trees, 19 Nectandra grandiflora, 13 Campomanesia xanthocarpa, 10 Cinnamomum vesiculosum, 10 Prunus brasiliensis, 9 Ocotea porosa and 8 Matayba elaeagnoides were assessed.

Mattos et al. (2007) analysed the dendrochronological potential of six species from the Mixed Ombrophilous Forest. Trunks were collected in Candói, PR, in an area belonging to ELEJOR, Centrais Elétricas do Rio Jordão, from three to six individuals of the following species: Araucaria angustifolia, Clethra scabra, Cedrela lilloi, Ocotea porosa, Podocarpus lambertii and Sebastiania commersoniana. The trees ranged from 14 cm to 40 cm in diameter at breast height (DBH) and were on average 60 years old, with an average annual increment of 0.6 cm. Although the number of trees was small, it was possible to observe that the environmental conditions were not limiting for the species, as the growth rings are not very sensitive. However, the extreme weather conditions experienced in 1999 and 2000, when there was a period of extremely low rainfall followed by a very harsh winter, affected the growth rings on many of the discs analysed in this study.

Araucaria angustifolia stood out when considering the periodic increase in DBH over the last 10 years ($IPA_{DBH(10)}$), with an average increase of 0.9 cm per year, ranging from 0.5 cm to 1.3 cm. Although Cedrela lilloi had an $IPA_{DAP(10)}$ of 0.6 cm, some individuals also showed periodic increases among the highest (1.1 cm and 1.3 cm). Clethra scabra was the slowest growing species (0.2 cm per year), while the others showed average periodic growth of 0.5 cm per year (MATTOS et al., 2007).

3.4 SOIL-PLANT RELATIONSHIP

Andrade and Sousa (1995) emphasise that the soil is of great importance in biological processes, as it is the physical environment in which a great variety of living beings live, whose life cycles contribute to the processes of soil formation, as well as anthropogenic actions, which contribute to the establishment of environmental changes. Therefore, the dynamics of nutrients in the body of the soil and in the physiological system of plants is influenced by all this complexity.

The chemical composition of plants takes in elements from the air (carbon), water (hydrogen and oxygen) and the other mineral elements are absorbed from the soil (FAQUIN, 1994).

The behaviour of plants, whether they belong to natural vegetation or man-made plantations, depends on a series of direct factors linked to the quality of the environment, as well as indirect factors (REZENDE et al., 1988).

Nutrients are one of these qualities, which are strongly interdependent with the morphological characteristics and properties of the soil. The interrelationships of dependence between nutrients and indirect factors (soil, vegetation, climate), as well as the interrelationships between nutrients and other direct factors, such as solar radiation, water, temperature, soil aeration and erosion, for example, illustrate some of the ramifications of this network of relationships (RESENDE et al., 1988).

The need to determine critical nutrient levels for each plant species or group of related species is based not only on their varying nutritional requirements, but also on their different nutrient absorption and/or utilisation efficiencies (BARROS and NOVAIS, 1990; ABICHEQUER and BOHNER, 1998).

Resende et al. (1988) emphasise that in situ assessment is fundamental for interpreting fertility conditions and nutrient availability, since the soil is part of a landscape that determines certain morphological characteristics, evidenced by variations in slope, climatic conditions, source materials and the influence of flora and fauna.

An important characteristic of plants is their ability to convert chemical elements absorbed from the environment into cellular components. For example, carbon, hydrogen, oxygen, nitrogen, phosphorus and sulphur are the elements that form proteins - protoplasmic constituents of vital importance for plant growth and development (VALE et al., 1994).

Plant nutrients are those chemical elements that are essential for plant production and without which plants cannot complete their life cycle. With the exception of carbon, oxygen and hydrogen, which are supplied to plants through water and air, the other essential elements are supplied through the soil. Therefore, fertility means the study of aspects related to the dynamics, supply and availability of plant nutrients (VALE et al., 1994).

The nutrients supplied by the soil are not demanded in equal quantities by the plants and the main factor controlling nutrient content is absorption potential, which is genetically fixed. For example, the foliar content of nitrogen and potassium is around ten times greater than that of potassium, sulphur and magnesium or around a thousand times greater than that of iron, zinc and manganese. These variations occur in all plant species. But between species, there can also be differences in the content of a given nutrient, which is also genetically determined (VALE et al., 1994).

The classic division classifies nutrients that occur in higher levels in plants as macronutrients (nitrogen, phosphorus, potassium, calcium, magnesium and sulphur). In turn, the least demanded nutrients are called micronutrients (iron, zinc, manganese, copper, molybdenum, boron and chlorine) (VALE et al., 1994).

According to Faquin (1994), the main factors (internal and external) that affect root ion absorption are: availability, pH, aeration, temperature, humidity, the element itself and interaction between ions. In this order, the first condition for the ion to be absorbed is that it is in available form and in contact with the root. Therefore, all the factors that affect availability also affect absorption.

In the study of plant nutrients, nitrogen stands out from the rest because it is highly dynamic in the soil system and is usually the nutrient most demanded by crops. This nutrient has complex dynamics, characterised by high mobility in the soil and various transformations in reactions mediated by microorganisms. Due to this dynamism, nitrogen, when compared to other nutrients, is much more difficult to keep in the soil within reach of the roots (VALE et al., 1994).

Unlike other nutrients, nitrogen is practically not supplied to the soil by source rocks, as the primary source is N_2 gas, which makes up 78 per cent of the Earth's atmosphere. However, nitrogen is not absorbed by plants in its elemental form and needs to be transformed into organic or inorganic forms that can be utilised by plants, as in the case of the transformation from mineral to organic nitrogen. The transformation of mineral nitrogen into organic nitrogen is called immobilisation and depends on the C/N ratio of the decomposing organic material (VALE et al., 1994).

In practice, when determining the capacity to supply nitrogen or produce a deficit of this nutrient in the soil after incorporating organic waste, a C/N ratio of around 20:1 has been considered the dividing line between immobilisation and mineralisation (VALE et al., 1994).

Soil is the habitat characteristic that most influences plant growth and its main attributes include texture, structure, temperature, pH, fertility, humidity and those related to the source material. Soil is therefore a decisive factor in the performance of tree species, whether planted in the open, under cover or in natural conditions (PRICHETT, 1979).

Santos Filho and Rocha (1987) emphasise the influence of soils on the productivity of tropical forests, as in a case study in the region of Lapa - PR, where the greatest growth of Araucaria angustifolia (Bert.) Ktze was found, related to the greater thickness of the solum and the A horizon. In the municipality of Telêmaco Borba - PR, they also mention the greater growth of Pinus taeda L. in sandy soils, on terraces in the lower parts of the natural landscape, and the lower growth in the higher part, where there is greater leaching and surface run-off.

De Hoogh (1981) emphasised the importance of soil characteristics as determining factors in the growth of Araucaria, since the species has a high nutritional demand.

Lopes (1983), citing work carried out by Haag et al. (1978), points out that forest essences behave differently from agricultural crops, as they contribute to improving the physical and chemical conditions of the soil in which they are planted, because their roots reach deeper and remove nutrients from the lower layers, which together with others absorbed by non-root pathways, form plant tissues, which are subsequently incorporated into the upper layers, producing material that is constantly transformed into humus by biological processes.

Galvão et al. (1999) state that plant succession is a dynamic process that involves forest phytoecological units undergoing accentuated physiognomic, structural and floristic changes when there is a transition from an initial, simpler phase to a more advanced phase of greater complexity and stability.

Wisniewski (1997) and Galvão (1999) emphasise that plant succession is a process of substitution, which requires changes in the amount of biomass stored in each phase, promoting an improvement in the soil's physical and chemical characteristics through the addition of organic matter.

Martinelli et al. (1999) emphasise that tropical forests are generally richer in nitrogen than temperate forests, due to the greater cycling of this element. He also adds that the intense solar radiation and high rainfall in tropical climate regions mean that soil organic matter decomposition rates are high, implying accelerated nutrient cycling and intense microbial activity.

Balbinot (2009) also cites organic matter as an important source of nitrogen in the soil, because due to biological activity the soil starts to contain two important elements that do not exist in the soil's original material: carbon and nitrogen. Nitrogen comes from small annual additions of inorganic nitrogen by rainwater and fixation of atmospheric nitrogen by microorganisms.

Souza (2006) found a $NO3^-$ input of 5.7 kg ha year $^{-1}$ in areas of Lowland Ombrophilous Dense Forest on the coast of Paraná, in secondary forests in the middle and advanced stages of regeneration. Considering that part of the carbon in the original plant material was transformed into carbon dioxide, with subsequent loss, the humus is enriched with nitrogen compared to the original plant material. It should be noted that the total carbon and nitrogen content is reduced by soil cultivation when compared to native forest. The litter is the main source of carbon for the soil.

According to Longman and Jeník (1987) the death of each organism in the forest ecosystem is followed by the decomposition of biomass into simpler organic and inorganic substances, releasing some of the energy they possess. Biomass production rates can be higher in tropical forests than in temperate forests, especially in the formation of foliage and shoots in the apical meristems, and also, under favourable conditions, in the production of wood. This is due to the different environmental conditions (temperature and humidity), which imply greater or lesser decomposition of the available material. In other words, the higher the temperature and humidity, the greater the activity of decomposer organisms, because as the author points out, fungi make a significant contribution to the nutrient cycle in tropical forests.

3. 5 LITTER

Leaf litter is made up of leaves, branches, bark and reproductive structures (flowers, fruit and seeds) deposited on the soil surface, with leaves being the main component (GOLLEY et al., 1978).

Forests are the largest accumulators of biomass on the planet, which is why there is a growing demand for information and knowledge that can help reduce environmental risks. Strictly speaking, biomass means the mass of matter of biological origin, living or dead, animal or vegetable. The term forest biomass can mean all the biomass in the forest or just its tree fraction (SANQUETTA, 2002).

The term phytomass has also been used to specify that it is biomass of plant origin. In this case, to better specify that it is the tree portion of phytomass, the term forest phytomass or tree phytomass could be used (SANQUETA, 2002).

Lopes (1983) states that the main means of returning minerals to the soil is through the decomposition of litter and, therefore, soil fertility is influenced by the speed of decomposition of this material, which is so important in terms of quantity and composition. The author emphasises that Pinus plantations less than 20 years old preserve many characteristics from the time they were planted. However, in older stands, the soils are improved due to the enrichment of nutrients by the prolonged deposition of leaf litter and the washing away of soluble salts from the canopy. The paper also mentions the effects of planting Pinus taeda in degraded areas at the age of 15, where the nitrogen and calcium levels were higher in the Pinus litter than in the adjacent herbaceous vegetation.

The study by Lopes (1983) also provides very relevant information about the influence of planting Pinus elliottii Engelm and other conifers on the physical and chemical conditions of the litter and the soil. The occurrence of low pH in soils under conifers is notorious, as is the lower removal of bases from the soil, especially calcium and, obviously, the lower return of this nutrient to the soil. A lowering of pH was observed under Pinus elliottii Engelm at 32 years of age, in areas previously occupied by Eucalyptus, with pH being the only property studied considerably affected by the change in vegetation.

Lopes (1983) reports that in the states of Paraná and Santa Catarina there has been a decrease in pH and an increase in exchangeable aluminium levels due to the replacement of native forest, mainly by Pinus elliottii Engelm. Organic matter in itself is an important indicator of fertility, as it is related to cation exchange capacity (CEC), nitrogen levels and other nutrients, but organic matter does not always decrease after planting with various species of Pinus, as the opposite results have been seen, where organic matter increased. However, some authors have observed that reforestation with pine trees caused a decrease in the levels of organic matter, nitrogen and carbon, as well as an increase in the carbon/nitrogen ratio (C/N).

Tosin (1977) observed a decrease in the levels of these exchangeable bases when replacing native forest with Pinus elliottii Engelm, and the same was found by Lepsch (1980) in plantations of Pinus in cerrado areas.

At the end of his work, Lopes (1983) emphasised that Pinus spp stands are young and produce little litter up to the age of 19. According to Florence and Lamb (1974), up to this age the soil is the main source of nutrients for the development of the plantation and only after the age of 20 does it become enriched in nutrients, due to the greater deposition of litter. The same study found that up to the age of 19, planting with pine trees in areas of natural cerrado vegetation caused an increase in aluminium, carbon, phosphorus, calcium, sum of bases, CTC, C/N ratio and a decrease in pH. The changes were small, considering that cerrado soils are already very acidic and have low natural fertility. It was therefore to be expected that the changes would not be great.

In another study carried out by Caldeira et al. (2008), which included the Parque das Nascentes in the municipality of Blumenau/SC, an experiment was carried out on quantifying litter and nutrients in a Dense Ombrophilous Forest, characterised by three successional stages, which were analysed separately. It was found that the average stock of accumulated litter ranged from 4.47 to 5.28 Mg ha^{-1}. The results obtained by Caldeira were within the range indicated by other authors in studies of forest formations (O'CONNELL and SANKARAN, 1997).

O'Connell and Sankaran (1997) point out that in certain locations in South America the accumulated litter production of natural tropical forests varies between 3.1 and 16.5 Mg ha^{-1}, with the maximum value (16.5 Mg $ha^{-1)}$ being observed in submontane forests in Colombia. According to Tanner (1980), montane forests generally accumulate more litter than other natural tropical forests, probably due to the low nutrient content in the leaves and mainly due to climatic factors, which imply slow deposition.

Still with regard to litter production, in a study carried out by Morellato (1992) in semi-deciduous forests in the south-east of Brazil, the values found ranged from 5.5 to 8.6 Mg ha^{-1}.

Another study comparing the deposition of organic matter in the soil, carried out by Garay et al. (2003) between two different areas planted with Eucalyptus grandis and Acacia mangium in

Espírito Santo, showed that the area planted with Acacia mangium had a greater accumulation of leaf litter (10 Mg ha^{-1} on average) than the area planted with Eucalyptus grandis (5 Mg ha^{-1} on average). The results indicate that there is greater incorporation of organic matter and nutrients in the soil under Acacia mangium compared to Eucalyptus grandis.

The accumulation of litter varies depending on the origin, species, forest cover, successional stage, age, time of collection, type of forest and location. In addition to these factors, others such as: soil and climate conditions and water regime, climatic conditions, site, understorey, silvicultural management, canopy proportion, as well as decomposition rate and natural disturbances such as fire and insect attack or artificial disturbances such as removal of litter and cultivation, occurring in the forest or stand, also influence litter accumulation (CUNHA, 1997).

This relationship between litterfall and successional stage was highlighted by Cunha (1997) in a study of accumulated litter biomass in a seasonal forest in Rio Grande do Sul, with different successional stages: capoeira at 13 years old, capoeirão at 19 years old and secondary forest over 30 years old, in which the following values were found: 4.2; 5.6 and 6.0 Mg. ha^{-1}, respectively. Brun et al. (2001) also quantified the biomass of litter accumulated in different successional stages in the Seasonal Deciduous Forest in Rio Grande do Sul, where they found the following results: capoeirão (5.1 Mg ha^{-1}), secondary forest (5.7 Mg ha^{-1}) and mature forest (7.1 Mg $ha^{(-1)}$).

Accumulated leaf litter is the main way of transferring nitrogen, potassium and calcium to the soil. This fact was highlighted by Caldeira (2003) in his study of the Montane Mixed Ombrophilous Forest in Paraná, which clearly shows the importance of the material that forms the litter in the process of biogeochemical cycling of nutrients in forest sites, especially for nitrogen, potassium and calcium.

According to Caldeira (2003), different levels and contents of macronutrients in the accumulated litter may be related to the mobility of bioelements within the plant, for example potassium, making it subject to leaching. According to Neves (2000); Pagano and Durigan (2000), the high variability of potassium levels in litter may be related to the variation in rainfall between the assessment periods. This can be explained by its high susceptibility to leaching via leaf and litter washing, which results from the fact that potassium does not participate in organic compounds, but occurs in soluble form or adsorbed in the cell juice (MARSCHNER, 1997).

Considering only the macronutrients, Ca has the second highest content in the accumulated leaf litter (CALDEIRA, 2003), a fact that may be related to its low mobility in plant tissues and the longevity of the leaves. The low mobility of this macronutrient within plant tissues is cited by Nilsson et al. (1995) as a factor that determines that the greatest amount of cycling of this nutrient in nature is done by the fall and decomposition of senescent plant tissues. The low levels of potassium in the accumulated litter are related to the low rates of this nutrient in biogeochemical cycling, unlike those of calcium, a nutrient whose levels in the accumulated litter are many times higher than those in the above-ground biomass components. Biogeochemical cycling, in general, is the way in which nutrients with low mobility in the plant are cycled, since biogeochemical cycling is not very significant for these nutrients, contrary to what happens for nutrients with high mobility in the plant (CALDEIRA, 2003).

Regarding micronutrients, regardless of the successional stage, the accumulated litter is the main route for transferring Fe, Mn and Zn to the soil. Both in this forest and in a Montane Mixed Ombrophilous Forest in Paraná (CALDEIRA, 2003), the accumulated litter in the three stages showed the following decreasing order of micronutrients: iron > manganese > zinc > boron > copper. The greater presence of iron may be due to its low mobility in plant tissues, according to Malavolta (2006). Low mobility is affected by high phosphorus content, potassium deficiency, high manganese content and low light intensity (DECHEN and NACHTIGALL, 2006). Caldeira (2003) points out that there may also be interference due to higher levels in the old leaves of certain species, as well as higher average levels in the leaves compared to the wood, bark and branches. Iron levels can also be influenced by soil contamination, as clays tend to retain iron and adequate levels of organic matter provide better utilisation of this micronutrient. Organic matter has acidifying and reducing characteristics and certain humic substances can form chelates under adverse pH conditions (DECHEN and NACHTIGALL, 2006). The iron content increases with increasing acidity, as well as in soils rich in humic acids and colloids capable of forming soluble complexes with iron (MALAVOLTA, 2006). The author also emphasises that the iron content in the soil is a consequence of the content in the source material.

The second micronutrient with the highest content in leaf litter is manganese, perhaps due to contamination with the soil, as this nutrient comes from oxides, carbonates, silicates and sulphides. Old leaves store more manganese when there is a good supply, according to Heenan and Campbell (1980), but there is also an influence of species and vegetation period (CALDEIRA, 2003) and (DECHEN and NACHTIGALL, 2006).

The same emphasis is placed on nutrient transfer by Rodriguez Jiménez (1988) in a study carried out in a lowland rain forest in Colombia. Schumacher et al. (2002) also observed this trend in plantations of

Araucaria angustifolia at 14 years of age, where the main transfer route for iron, manganese and zinc to the soil was the accumulated litter. Boron is the fourth most important micronutrient in accumulated litter, due to its low mobility in plant tissues.

Caldeira et al. (2008) emphasise that leaf litter is also the main way of transferring organic carbon to the soil, mainly through the fall of senescent components from the aerial part of the canopy, which is why it is very important to quantify it. However, it is essential to emphasise that roots, dead wood, micro, meso and macro fauna are also important ways of transferring organic carbon to the soil.

Accumulated leaf litter plays an essential role in plant growth, as it influences the physical, biological and chemical properties of soils, also contributing to an increase in the soil's cation exchange capacity (CEC) (SYERS and CRASWEL, 1995; GARAY; ANDRADE and KINDEL, 2001).

Le Bourlegat et al. (2007) carried out interesting research into the structure and mass of litter in reforestations of different ages and forest fragments in northern Paraná. One of the objectives of the study was to compare the litter necromass of reforestations of different ages and to compare it with adjacent forest fragments, in order to assess the incorporation of biomass in the development of reforestations, since the increase in forest biomass is correlated with the capture of carbon from the atmosphere, since the organic matter stored in a forest ecosystem is directly represented by its biomass. In addition to the carbon fixed to the living mass of a forest, there is also that incorporated into the abiotic system, such as the soil. The authors also emphasise that despite this importance, studies of biomass accumulation are one of the little-known aspects of forest ecosystems (BURGER and DELLITTI, 1999).

The results obtained by Le Bourlegat et al. (2007) show that reforestation for ecological restoration purposes is an important source of carbon capture from the atmosphere and can present results comparable to those of forest fragments in terms of organic matter deposition on the soil Balbinot et al. (2003).

Wisniewski and Reissmann (1996), studying litter deposition in Pinus taeda L. plantations in the region of Ponta Grossa - PR, found values varying between 6 and 8 Mg ha year $^{-1}$, in line with what was predicted by Bray and Gorham (1964).

3.6 CARBON IN THE SOIL

The distribution of organic carbon in soil profiles varies greatly on a global level, with lower amounts stored on the surface of tropical soils than in forest soils at lower latitudes. On average, according to Schlesinger (1977), 1 per cent of the carbon in a given soil profile is stored in the top layer of litter in tropical forests, while boreal forests store 13 per cent.

Forests represent an important alternative source of energy, as they are a renewable natural resource and also contribute to reducing the environmental impacts of the greenhouse effect and its implications for climate change, due to their potential to retain carbon in the atmosphere (SANQUETTA, 2002).

In view of the above, forest biomass has a decisive influence because it is a more rational energy source and also because it accumulates pollutants in its carbon structure that are harmful to the quality of life on the planet. For this reason, interest in research into forest biomass and carbon content is growing all the time, motivating many scientists and institutions to expand their analyses of the subject (SANQUETTA, 2002).

It is estimated that the total carbon stock exceeds 26.10^{15} Mg, most of which is in the form of inorganic compounds and only around 0.05% in organic form. Organic compounds are found in marine and terrestrial biomass, organic debris and soil (terrestrial), as well as in the sediments and organic debris of the oceans (LARCHER, 2000). Of the total carbon on Earth, only 0.05 per cent is made up of organic compounds, mostly marine debris and sediments and soil. Within the phytomass, forests play a leading role, storing more than % of the carbon found in terrestrial plants (LARCHER, 2000).

Rochadelli (2001) emphasises that the conservation of natural and planted forests is of paramount importance for reducing carbon dioxide levels in the atmosphere, since plants absorb carbon dioxide and convert it into carbohydrates in the form of wood tissue, leaves, seeds and fruit. During their evolutionary cycle, these forests capture and fix carbon in their wood and other biomass components. Quantifying organic carbon is therefore important because during the existence of these forests, quantities of carbon in the atmosphere are immobilised, especially in the wood.

According to some important considerations by SCHUMACHER et al. (2002), when discussing the subject of tree biomass and organic carbon, it is possible to point to the planting of forest species in degraded areas, with the aim of enabling the use of agroforestry systems, which help to fix carbon in the soil.

Despite the lower carbon stock in forest biomass, when compared to the atmosphere, soil, fossil fuels and ocean compartments, this is the only one that can be easily managed for carbon sequestration purposes. Utilising the biomass of fast-growing forests as an energy source instead of fossil fuels would be an important alternative for mitigating the effects of carbon increase in the atmosphere (SANQUETTA, 2002).

Gardner and Mankin (1981) state that forest ecosystems contain around 90 per cent of the Earth's biomass, covering approximately 40 per cent of its surface. According to Kozlowski and Pallardy (1996), the biomass accumulated in ecosystems is affected by all those factors related to photosynthesis and respiration. In this sense, Campos (1991) and Caldeira (1998) add that this accumulation (production) is different from place to place where it is measured, reflecting the variation of the various environmental factors and factors inherent to the plant itself. Different types of forest store different amounts of carbon within their biomass, and different sites within the same type of forest also vary greatly in terms of the amount of biomass.

Schlesinger (1977) reports that at a global level, the average amount of organic carbon in forest soil increases from the tropics to temperate to boreal forests. Low soil organic carbon values in tropical forests are caused by rapid decomposition, which compensates for the rapid production of litter. The variation between tropical forests can be very large, reaching comparable differences between tropical and temperate and boreal forests (ANDERSON and SWIFT, 1983).

In one of his studies, Schumacher (2000) found an average above-ground biomass of 2.84 Mg ha^{-1} in native grassland vegetation in the north-eastern region of the state of Rio Grande do Sul, which corresponded to a carbon stock of 0.99 Mg ha^{-1}. This value corresponds to less than 1% of the carbon fixed in the biomass of Pinus taeda L., 20 years old.

With regard to soil organic carbon concentrations, Mafra et al. (2008) carried out a comparative study of organic carbon levels and soil chemical attributes between different sites: native grassland, pine plantations (12 and 20 years old respectively), araucaria reforestation (18 years old) and native araucaria forest. In this study, it was observed that on the Pinus and Araucaria sites the organic C stocks were maintained at levels equivalent to those in the forest and field, with values of 12.5 to

14.2 Kg m^{-2}, in the 0.0 - 0.4 m layers. Considering the average of the layers analysed, the acidity and available P contents were higher in the 20-year-old Pinus plantation. The chemical analysis revealed low to medium levels of available phosphorus and exchangeable potassium, calcium and magnesium.

Lal, Kimble and Follet (1998) state that the balance between the processes that increase and decrease the carbon stock in the soil is influenced by land use and anthropogenic factors, and that these factors determine the stock of the element in the pedosphere. The forest not only protects the soil against erosive processes, but above all contributes to the maintenance of the enormous carbon stock, due to the reduced impact on planting seasons, which would not occur if it were used for agriculture, which would imply exposure to intensive preparation and short periods of time, resulting in the loss of tonnes of CO_2 to the atmosphere due to accelerated decomposition and oxidation.

One of the factors influencing the concentration of carbon in the soil is soil density. According to Ferreira (1994), Soil Density (SD) is fundamental for quantifying organic carbon in the soil. Density is also known as Apparent Density and Global Density and represents the relationship between the mass of solids (Ms) and the total volume (Vt), i.e. the volume of the soil including the space occupied by water (V_{H2O}) and air (Va):

$$Ds = Ms / Vt$$

Ds is a physical property that reflects the arrangement of soil particles, which in turn defines the characteristics of the porous system. In this way, any manifestations that influence the arrangement of soil particles will directly influence the Ds values (FERREIRA, 1994).

CHAPTER 4

MATERIAL AND METHODS

4.1 DESCRIPTION AND LOCATION OF THE STUDY AREA

The Colégio Florestal Estadual Presidente Costa e Silva, the site of the research, is located in the municipality of Irati, in the southeastern mesoregion of Paraná (Figure 1), about 150 km from Curitiba, with an average altitude of 812 metres above sea level and coordinates of 25°28'S, 50°39'W. The Forestry College (Figure 2) has a total area of 176.5 ha.

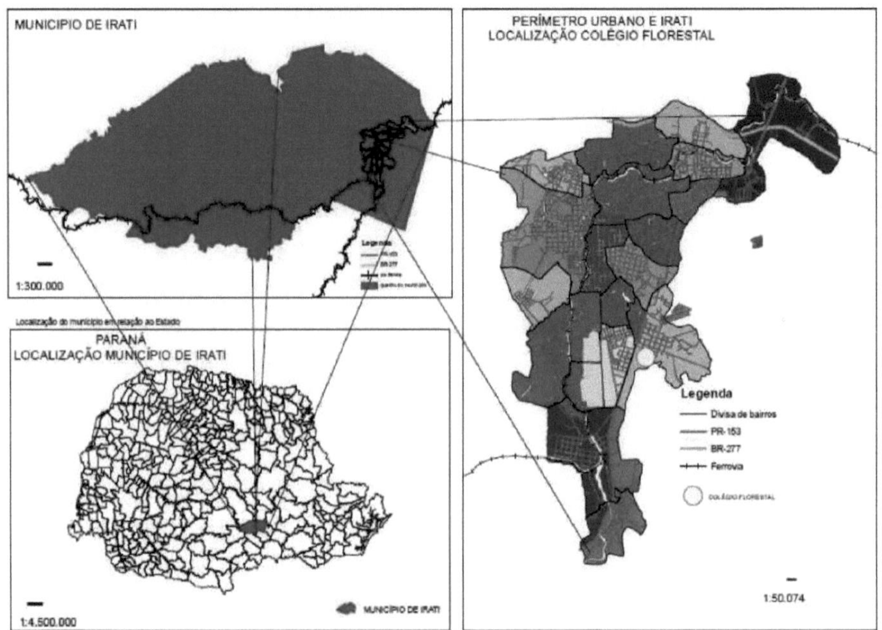

Figure 1 - Location of the Municipality of Irati and the Study Area.
Source: Geoprocessing Department of Irati City Hall, 2010.

Figure 2 - Aerial view of the Forestry College.
Source: OLIVEIRA, 2008.

According to the Koppen classification, the study area has a Cfb climate (humid subtropical, mesothermal, with cool summers and frequent severe frosts, average annual rainfall of 1200 mm).

With regard to the characteristics of the area studied, the Colégio Florestal Estadual Presidente Costa e Silva has an area of 176.5 hectares, of which 60 hectares are covered by native forest vegetation, classified as a fragment of Montane Mixed Ombrophilous Forest (Figures 3 and 4), kept in a relatively preserved state. In this forest formation, typical of the second plateau of Paraná, the second successional stage predominates, called intermediate, characterised by very dense tree vegetation, still dominated by pioneer and early secondary species, generally with a low diametric range.

The area of the Montane Ombrophilous Mixed Forest (FOMM) is quite shaded, with relatively dense vegetation and little degradation (Figure 3). In this area, there are also some springs, very humid soils and poor drainage conditions. The predominant soil is Cambissolo, developed on sedimentary material of the shale type.

According to Medeiros, (2008) (unpublished data), the floristic composition of the fragment of Montane Mixed Ombrophilous Forest in the Colégio Florestal area has common species of great diversity, distributed homogeneously. The total area

of native forest is approximately 60 ha, with around 171 tree and shrub species, including two unidentified species. It can be said that despite the great diversity of species, the area has been exploited in the past and is in the process of forest succession.

In terms of phytosociology, the following stand out: Pimenteira - *Cinnamodendron dinisii*

Schwacke (Cannelaceae), Jerivá - *Syagrus romanzoffiana* Glassman, Guaçatunga-red - Casearia obliqua Spreng (Salicaceae), Canela-amarela - Nectandra lanceolata Nees et Mart. (Lauraceae), Peach tree - Prunus brasiliensis Chamisso & Schlechtd (Rosaceae), Yerba mate - Ilex paraguariensis St. Hill (Aquifoliaceae). Hill (Aquifoliaceae), Curitiba prismatica (Myrtaceae), Guaçatunga-miúda - Casearia decandra Jacq (Salicaceae), and Mamica-de-cadela - Zanthoxylum rhoifolium Lam. (Rutaceae) (MEDEIROS , 2008).

According to Medeiros (2008), these data came from phytosociological analyses and various observations carried out between February 2000 and December 2008. In general, species richness was not very different throughout the area, except in the stretches of riparian forest, both along the Rio das Antas and other watercourses. The FOMM fragment was classified as being at a medium stage of ecological succession. According to Conama Resolution 06 (Brazil, 1994), which establishes measurable parameters for analysing stages of ecological succession, a medium stage of succession is defined as an average DBH of 10 - 20 cm, an average height of 5 - 12 m and an average basal area of 10 - 28 m^2 ha^{-1}.

The Pinus elliottii Engelm is a 33-year-old plantation (Figure 5), with an area of 4.2 ha, which has already been thinned three times. The spacing adopted was 2.0 x 2.5 $metres^2$. In the year of establishment (1977) there were 2,541.59 trees. The first systematic thinning was carried out in 1985, removing approximately 110 m^3, equivalent to 20% of the plot. The second thinning was carried out in 1989 and the third in 1997, both selective. In 2005, 508 trees remained after the third thinning, whose average annual increment (IMA) was 91.09 m^3 ha $year^{-1}$. Pruning was carried out on 2.1 ha, to a height of 2.0 m in 1982 and to a height of 4.0 m in 1984 (MANASSÉS, 2007 - unpublished data).

In this plot of Pinus elliottii Engelm, silvicultural practices were carried out to control leaf-cutting ants and the wood wasp. This area is used for practical forestry classes and is made up of relatively young soils with low natural fertility, acidic pH and thin, well-defined horizons. As it is a plantation area, the incidence of light changes according to the interventions in the forest.

Figure 3 - View of the area of the Montane Mixed Ombrophilous Forest site.
Source: OLIVEIRA, 2008.

Figure 4 - Detail of the Montane Mixed Ombrophilous Forest site.
Source: OLIVEIRA, 2009.

Figure 5 - View of the *Pinus elliottii* plantation site.
Source: OLIVEIRA, 2008.

4.2 SOILS

According to Embrapa (2006), the soil types where the FOMM and the *Pinus elliottii* plantation are located are classified as typical alumino HUMIC CAMBISSOLO and dystrophic LITHOLIC NEOSSOLO (Figures 7 and 8).

Figure 7 - Detail of the soil profile in *Pinus elliotti* Dystrophic LITHOLIC NEOSSOLO
Source: OLIVEIRA, 2009.
Figure 8 - Detail of the soil profile in FOMM CAMBISSOLO HÚMICO ALUMÍNICO Típico
Source: OLIVEIRA, 2009.

Table 1 gives a general description of the soil profiles under study.

Table 1 - Soil profile for FOMM and planting of *Pinus Elliottii* Engelm.

Classification	Typical aluminaic humic cambisol FOMM	LITHOLIC Dystrophic NEOSSOLO *Pinus Elliottii* Engelm
Institution	Forest College	Forest College
Municipality	Irati, PR	Irati, PR
Location	300 m to the west of the established toposequence, on the firebreak after the apiary.	Toposequence CD, 200 m above the nursery, plot 4
Altitude	905 m	880 m
Geological Formation	Sheets of the Passa Dois Group	Sheets of the Passa Dois Group
Relief	Gentle undulating to undulating	Gentle undulating to undulating
Drainage	Imperfectly drained	Well drained
Vegetation	FOMM	Planting *Pinus Elliottii* Engelm
Current use	Preservation and research	Forest planting

Source: Oliveira and De Hoogh (1990).

4.3 QUANTIFICATION OF ACCUMULATED LITTER

According to Péllico Netto and Brena (1997) it is important to emphasise that the sampling method refers to the configuration of the sampling unit, while the term sampling process refers to the way the sample is placed or distributed in the population (randomly, systematically, etc.). Direct methods imply determinations, while indirect methods generate estimates. Determinations are not possible in large tracts of forest, but small areas and samples taken from the population can be used to adjust and calibrate the models used to estimate biomass. It can therefore be said that most forest biomass studies generate estimates rather than determinations.

The accumulated litter (Figure 9) in the two selected areas (Montane Ombrophilous Mixed Forest and Pinus elliottii Engelm plantation) was collected in May 2009, using the methodology adopted by Sanqueta (2002), using a 0.25 m x 0.25 m (0.0625 $m^{2)}$ metal template. The tracts of land covered to collect the material were 2,000 m^2, corresponding to 20 m wide by 100 m long strips, both in the FOMM and in the Pinus elliottii Engelm plantation, where 30 randomly distributed litter samples were collected to analyse nutrient content and biomass.

Figure 9 - Collecting leaf litter. Source: OLIVEIRA, 2009.

To determine the dry weight of the accumulated litter samples, they were placed in paper bags, then dried in an air circulation oven at 65°C until they reached a constant dry mass, which was measured on a precision scale (0.01g). The amount of accumulated litter found on the metal template (g/0.0625 m^2) was extrapolated to Mg ha^{-1}, using the average mass in grams contained in the area of the template (0.0625 m^2), proportional to 10,000 m^2 (hectare). The samples were then ground in a Wiley-type mill, passed through 1.0 mm mesh sieves (20 mesh) and stored in glass jars for subsequent chemical analysis (TEDESCO et *al.,* 1995; MIYAZAWA et al., 1999).

4.4 NUTRIENTS AND ORGANIC CARBON IN THE ACCUMULATED LITTER

The 30 single samples collected in the two 2000 m^2 strips, in the sampling unit corresponding to the metal template (0.0625 m^2), in the two forest formations, were homogenised into 5 composite samples, from which approximately 3g of each composite sample was separated to be sent for analysis in the laboratory.

With regard to plant tissue for the determination of macronutrients, nitrogen was determined in the sulphuric digestion extract using the Kjeldahl method (distillation - titration). The other elements (phosphorus, potassium, calcium, magnesium and sulphur) were
determined in the nitric-perchloric digestion extract, phosphorus and sulphur by spectrophotometry (UV-VIS), potassium by flame photometry and calcium and magnesium by atomic absorption spectrophotometry (MIYAZAWA et al., 1999). Organic carbon was determined using the Walkley-Black method (TEDESCO et al., 1995; MIYAZAWA et al., 1999).

Analyses to determine the levels of nutrients contained in the litter were carried out by the

Environmental and Instrumental Chemistry Laboratory of the State University of Western Paraná - UNIOESTE.

To determine the C/N ratio, the CHN method, carried out by the Soil Department of the Federal University of Paraná, was used.

4.5 CARBON IN THE SOIL

The soil samples used to determine the carbon content and quantity were collected in mini-profiles at three different depths: 0 - 5 cm, 5 -15 cm and 15 - 30 cm. In each mini-profile, which was opened with a common cutter, the samples were collected with metal cylinders of known volume at the three depths. A total of 30 mini-profiles were opened at random along a 2,000 metre strip^2 in the FOMM and in the Pinus elliottii Engelm plantation. The samples were then dried in an oven and weighed on precision scales (0.01 g).

Soil density was calculated using the following expression (EMBRAPA, 2002):

$$\mathbf{DA = PSN / VCH}$$

In which:

DA = soil density

PSN = dry weight of the soil inside the cylinder

VCH = Uhland cylinder volume (constant)

In this study, the cylinder used (Figures 9 and 10) has dimensions of 5.5 cm in diameter and 2.90 cm in height. Therefore, the volume of the cylinder is 68.89 cm^3.

The Kopecky ring method was used to collect the soil for analysing soil density (EMBRAPA, 1997).

The following expression was used to calculate the carbon stock for a given depth:

$$\mathbf{Est\ C = (CO \times Ds \times e)/10}$$

In which:

Est C = stock of organic C at a given depth ($Mg.ha^{-1}$)

CO = total organic C content at the depth sampled ($g.kg^{-1}$)

Ds = soil density of the depth ($g.cm^{-3}$) **e** = thickness of the layer considered (cm)

1.6 CHEMICAL AND PHYSICAL ANALYSIS OF THE SOIL

Soil samples (taken in May 2009) were collected in the two selected areas (FOMM and Pinus Elliottii Engelm) using a cutting spade, with randomly distributed collection points. 30 samples were used per selected area, at the following depths: 0-5 cm; 5-15 cm and 15-30 cm.

The physical and chemical analyses used 30 single samples collected in plastic buckets at the three depths (0 - 5 cm; 5 - 15 cm; 15 - 30 cm). The samples were then homogenised to form composite samples by depth in the FOMM and in the Pinus elliottii Engelm plantation. The physical and chemical analyses were carried out at the Forest Soils Laboratory of the Forest Engineering Department/Unicentro, according to the methodology proposed by Tedesco et al. (1995).

The organic carbon content was calculated from the result of the MO (organic matter) content, according to the expression: MO = C x 1.72, which was carried out by the Forest Soils Laboratory of the Forest Engineering Department at Unicentro - Irati.

4. 7 STATISTICAL ANALYSIS

We used Sudent's t-test for significance, which is a test for comparing means for independent samples (STUDENT, 1908). The area of 2,000 m^2 adopted in this research for the FOMM and Pinus elliottii Engelm plantation represents the population, where the sampling units (metal templates) were distributed. The 30 litter samples were collected in these sampling units. This 2,000 m $strip^2$ is fairly representative of the FOMM, which shows homogeneity in the distribution of tree species and the predominant soil.

CHAPTER 5

RESULTS AND DISCUSSION

5.1 QUANTIFICATION OF ACCUMULATED LITTER AND STATISTICAL ANALYSIS.

With regard to litter production for Pinus elliottii Engelm and FOMM, respectively, the values were 12.89 Mg ha^{-1} and 12.61 Mg ha^{-1}.

The statistical analysis used the values obtained from the 30 samples (g), proportional to the area of the metal template ($0.0625m^2$) and then converted to hectares (Mg ha^{-1}).

The statistical analysis (Table 2) for independent samples in the two forest formations revealed that there was no significant difference between the amounts of accumulated litter at the 5% significance level. As the calculated "t" (0.2144) is lower than the tabulated "t", it can be concluded that, on average, the samples are statistically equal, i.e. there is no significant difference between the amounts of accumulated litter.

Table 2 - Descriptive data (mean, standard deviation, minimum and maximum) of accumulated litter and significance of Student's t-test for FOMM and Pinus elliottii Engelm.

	Pinus elliottii **Engelm**	Ombrophilous Forest ***Mixed***
Average	80,5823333	78,84
Variance	602,606584	1378,06169
Observations	30	30
Mean difference hypothesis	0	
gl	50	
Stat t	0,21443022	
P(T<=t) one-tailed	0,41554186	
one-tailed critical t	1,67590503	
P(T<=t) two-tailed	0,83108371	
two-tailed critical t	2,00855907	

Note: t-test: two samples assuming different variances

O'Connell and Sankaran (1997) point out that in certain locations in South America the accumulated litter production in natural tropical forests varies between 3.1 and 16.5 Mg ha^{-1}, with the maximum value (16.5 Mg $ha^{-1)}$being observed in submontane forests in Colombia. The values shown above for Pinus elliottii Engelm (12.89 Mg ha^{-1}) and FOMM (12.61 Mg ha^{-1}) are very consistent with this range of results. In comparison with the studies by Caldeira et al. (2008) in the Dense Ombrophylous Forest, the average stock of accumulated litter ranged from 4.47 Mg ha^{-1} to 5.28 Mg $ha^{(-1})$ and Caldeira et al. (2007) in the Montane Mixed Ombrophilous Forest, the accumulation of leaf litter was 7.99 Mg ha^{-1}, i.e. 2.8% in relation to the total biomass, which was 280.73 Mg ha^{-1}. These values are lower than those observed in five semi-deciduous forests in south-eastern Brazil, with values ranging from 5.5 Mg ha^{-1} to 8.6 Mg $ha^{(-1})$ (MORELLATO, 1992).

In a comparative study between reforestations planted with Semideciduous Seasonal Forest species of different ages (2.5, 3.0, 4.0 and 5.0 years) and fragments of Semideciduous Seasonal Forest in northern Paraná, Le Bourlegat et al. (2007) found average values of 6.0 Mg ha^{-1} for reforestations and 5.4 Mg ha^{-1} for fragments of the Semideciduous Seasonal Forest, which are lower than the results found in this study.

Comparing the results of the accumulated litter of the Montane Mixed Ombrophilous Forest under study (12.61 Mg ha^{-1}) with that of the Lower Montane Humid Tropical Forest, Rio de Janeiro, the latter has very high values (20.5 Mg $ha^{-1)}$. On the other hand, according to Clevelario Júnior (1996), high litter biomass values are associated with long development cycles, which are generally found in nutrient-poor forests such as the Low Montane Humid Tropical type. In forests with a greater scarcity of nutrients, the accumulation of litter is associated with the slowing down of decomposition and not with the more intense fall of forming material. In forests with greater nutrient scarcity, there is a greater accumulation of litter, reducing the speed with which nutrients are cycled.

In the Ombrophilous Forests of Australia, accumulated litter can vary from 4.4 to 6.3 Mg ha^{-1} (SPAIN, 1984). According to a review by O'Connell and Sankaran (1997), more than 75% of the accumulated litter in natural rainforests is below 7.0 Mg ha^{-1}, with an average of 6.0 Mg ha^{-1}.

Kleinpaul et al. (2005), studying the accumulated litter on the ground in Pinus elliottii Engelm (21 years old), Eucalyptus sp. (12 years old) and Seasonal Deciduous Forest, obtained values of 12.53 Mg ha^{-1}, in the acicles item, in Pinus elliottii Engelm, 4.17 Mg $ha^{(-1})$ in Eucalyptus and 4.0 Mg $ha^{(-1})$ in Seasonal Deciduous Forest. Using the Tukey test at a 5% probability of error, the statistical analysis revealed a significant difference in the accumulation of litter in Pinus elliottii Engelm in relation to Eucalyptus and the Seasonal Deciduous Forest, while the values found in the Eucalyptus sp. and Seasonal Deciduous Forest stands did not differ from each other.

Investigating a comparative analysis of litter accumulation in native forest and commercial reforestation of Pinus taeda L. (38 years old) and Eucalyptus grandis (also 38 years old), Medri et al. (2009) found the highest litter accumulation in Pinus taeda L. reforestation (11.7 Mg ha^{-1}), followed by Eucalyptus grandis (10.2 Mg $ha^{(-1})$) and native forest (5.8 Mg $ha^{(-1})$). The average stock of accumulated litter in the native forest was similar to that observed in five semi-deciduous forests in south-eastern Brazil, with values ranging from 5.5 Mg ha^{-1} to 8.6 Mg $ha^{(-1)}$. It should be noted that the high values found in the reforestations may be due to the behaviour and morphology of the species. It is important to note that the 52,000 ha area of native forest in this study has different vegetation formations: Semideciduous Seasonal Forest, Mixed Ombrophilous Forest and small patches of grassland.

As for the variation in the amount of litter accumulated in forest soils, there are several factors that interfere, including: unfavourable conditions for decomposition, the physical and chemical properties and content of the litter, the low density of decomposer organisms and the time of accumulated collection, as well as pruning during the sampling period, which will influence the contribution of material.

The variation pointed out by Balbinot (2003) occurred in an inventory of organic carbon in a Pinus taeda plantation, at 5 years of age, in Rio Grande do Sul, in which they estimated organic carbon from biomass, soil and litter. As for the litter on the ground, this was estimated at 17.4 Mg ha $^{-1}$, a high value due to the silvicultural practice of pruning.

The accumulation of litter on the soil surface is regulated by the amount of material that falls from the aerial part of the plants and its rate of decomposition. According to Pritchett (1990), this accumulation is due to the annual amount of litter minus the annual rate of decomposition. The rate of litter fall is uniform among tree species growing in similar climate and soil conditions.

The accumulation of litter varies depending on the origin, species, forest cover, successional stage, age, time of collection, type of forest and location. In addition to these factors, others such as: soil and climate conditions and water regime, climatic conditions, site, understorey, silvicultural management, canopy proportion, as well as decomposition rate and natural disturbances such as fire and insect attack or artificial disturbances such as removal of litter and cultivation, occurring in the forest or stand, also influence the accumulation of litter (CALDEIRA et al., 2007; CALDEIRA et al., 2008) Regardless of the factors that caused the differences in accumulated litter between Pinus elliottii Engelm and FOMM, the processes of falling leaves, branches, flowers, fruit, etc., improve the fertility of the topsoil (MAGALHÃES and BLUM, 1999), establishing a C/N ratio and a pH more favourable to biotic development (NOVAIS and POGGIANI, 1983).

When comparing the data obtained in this study with Watzlawick's (2003) data for Pinus taeda L. stands, a certain similarity can be seen. The author found 10.63 Mg ha^{-1} of accumulated litter in Pinus taeda L. stands, aged 32 years. Watzlawick's (2003) research also revealed results of 5.78 Mg ha^{-1} of accumulated litter in stands of Araucaria angustifolia (Bert.) Ktze, at 32 years of age.

In this study, the accumulated litter values for Pinus elliottii Engelm, (12.89 Mg ha^{-1}) were very similar to those of Pinus taeda L. (10.63 Mg ha^{-1}), at similar ages, 33 and 32 years, respectively, and approximately double the value found for Araucaria angustifolia (Bert.) Ktze (5.78 Mg ha^{-1}), in relation to FOMM (12.61 Mg ha$^{(-1)}$).

In another study by Watzlawick and Caldeira (2004), the average amount of accumulated litter was 8.01 Mg ha^{-1}, and the amount of organic carbon was 3.06 Mg ha^{-1}, in a Montane Mixed Ombrophilous Forest. In the same forest formation, in Irati (PR) 2009, the result was 12.61 Mg ha^{-1}, both within the minimum and maximum ranges most commonly found in related studies.

5. 2 STATISTICAL ANALYSIS OF NUTRIENTS IN ACCUMULATED LITTER

The averages obtained (Table 3) in the different forest types (Pinus and FOMM) were compared using Student's t-test (a=0.05).

The statistical analysis applied in this research, as shown in Table 3, revealed significant differences between the Pinus elliottii Engelm and FOMM stands in terms of nitrogen, potassium, calcium and magnesium levels. Therefore, the most significant differences were in macronutrients, due to the greater nutrient cycling capacity of tropical forests compared to planted forests and also due to the mobility of bioelements within the plant.

Table 3 - Descriptive data (mean, standard deviation, minimum and maximum) for nutrients and the significance of Student's t-test for Montane Ombrophylous Mixed Forest and Pinus elliottii.

Nutrient		Average	Standard deviation	Minimum	Maximum	t-test	P
			Macronutrient				
N	Pine	6,47	1,17	5,25	7,88	5,522 *	0,003
N	FOMM	14,70	3,11	9,63	17,5		
P	Pine	0,50	0,13	0,34	0,72	2.079 ns	0,106
P	FOMM	1,08	0,60	0,68	2,15		
K	Pine	1,05	0,19	0,85	1,25	3,276* *	0,022
K	FOMM	2,01	0,62	1,40	3,00		
Ca	Pine	14,36	2,19	11,40	16,60	2,326*	0,048
Ca	FOMM	17,69	2,32	15,10	20,30		
Mg	Pine	2,65	0,90	1,25	3,70	-5,203*	0,001
Mg	FOMM	4,96	0,40	4,60	5,60		
			Micronutrient				
Fe	Pine	823,00	367,90	470,00	1405,00	-0,227"	0,826
Fe	FOMM	864,00	165,20	610,00	53,00		
Zn	Pine	1710,00	1046,99	815,00	3470,00	1.672 ns	0,133
Zn	FOMM	2917,00	1228,95	1510,00	4150,00		
Ass	Pine	5,40	1,34	4,00	7,00	2.382 ns	0,076
Ass	FOMM	20,60	14,20	7,00	44,00		
Mn	Pine	29,60	5,27	24,00	37,00	-4,781* *	0,001
Mn	FOMM	45,40	5,17	40,00	53,00		

* significant at 5 % significance level.

As for the micronutrients, only manganese showed significant differences between Pinus elliottii Engelm and FOMM, considering that the higher manganese levels and contents in the accumulated litter may have occurred due to the higher levels in the leaves of some species, and also because the concentrations of this micronutrient in the leaves increase as the plant ages.

5. 3 SAMPLING THE ACCUMULATED LITTER

The average accumulated litter of Pinus elliottii Engelm (Table 4) was 75.24 g/0.0625 m^2, with a confidence interval at the 95% probability level showing a lower limit of 65.5 and an upper limit of 84.95. The coefficient of variation found was 27.19 %, and the number of samples needed to obtain an error of 10 % of the mean with 95 % probability for Pinus elliottii Engelm was 28 samples and 7 samples for an error of 20 % of the mean with the same probability. The true average of the accumulated litter in the forest under study was estimated with a 9.72 % error at the 95% probability level.

In the Montane Mixed Ombrophilous Forest (Table 4), the average accumulated litter was 70.21 g/0.0625 m^2, with a confidence interval at the 95% probability level showing a lower limit of 60.49 and an upper limit of 79.92. The coefficient of variation found was 39.44 %, and the number of samples needed to obtain an error of 10 % of the mean with 95 % probability was 59 samples and 15 samples for an error of 20 % of the mean with the same probability. The true average of the accumulated litter in the forest under study was estimated with a 14.11 % error at the 95% probability level.

Therefore, the sampling carried out was sufficient and representative of the area adopted, but more samples would be needed for FOMM, for a 10 % error in sampling intensity.

Table 4 - Statistical variables of accumulated litterfall sampling in Pinus elliottii and Montane Mixed Ombrophylous Forest.

Statistical variables	Pinus elliottii **Engelm**	**Mixed Ombrophilous Forest Montana**
Average (x)	75.24 g /0.0625 m^2	70.21 g/0.0625 m^2
Variance (s^2)	418,54	766,94
Standard deviation (s)	20,45	27,69
Standard error (s_x)	3,73	5,05
Absolute sampling error	± 7,317	± 9,908
Relative sampling error	9,72 %	14,11 %
Confidence interval ($\alpha = 5\%$)	IC95 $[65.5 \leq \mu \leq 84.95] = 0.95$	IC95 $[60.49 \leq \mu \leq 79.92] = 0.95$
Coefficient of variation (CV)	27,19 %	39,44 %
Sampling intensity (10% error) of the mean and 95% probability $n = t^2 * \frac{s^2}{E^2}$	28 samples per sampling unit	59 samples per sampling unit
Sampling intensity (20% error) of the mean and 95% probability $n = t^2 * \frac{s^2}{E^2}$	7 samples per sampling unit	15 samples per sampling unit

5.4 COMPARISON OF ACCUMULATED LITTER IN DIFFERENT FORESTS

The accumulation of litter on forest soils is determined by various factors, such as the type of forest, canopy formation, forest regeneration, species diversity, climatic agents, water regime, time of year, type and concentration of microorganisms, characteristics of the deposited material, fires, pest attacks and anthropogenic interventions, among others.

When comparing the amounts of litter accumulated in different forests (Figure 11), it can be seen that the values found vary within a fairly well-known range of results - 3.1 to 16.5 Mg ha^{-1} (O'CONNELL and SANKARAN, 1997). In this study, the values found in Pinus elliottii Engelm. and Montane Mixed Ombrophilous Forest (FOMM Irati) were very similar (12.89 Mg ha^{-1} and 12.61 Mg ha^{-1} respectively), which indicates the importance of both natural and planted forests for accumulating litter and, consequently, carbon and nutrients.

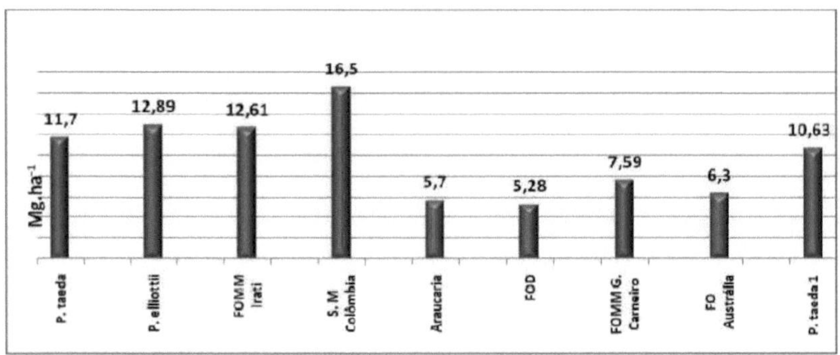

Figure 11 - Histogram comparing accumulated litter in different regions.

The references for the forest formations represented in the histogram are respectively: Pinus taeda L. (MEDRI, 2009), Pinus elliottii Engelm (OLIVEIRA, 2009), FOMM Irati (OLIVEIRA, 2009), Submontane Forest Colombia (O'CONNELL and SANKARAN, 1997), Araucaria angustifolia (Bert) O. Ktze (WATZLAWICK, 2003), FOD (CALDEIRA et al., 2008), FOMM Gal. Carneiro, PR (CALDEIRA, 2003), FO Australia (SPAIN, 1984) and Pinus taeda L. (WATZLAWICK, 2003).

5.5 NUTRIENTS AND ORGANIC CARBON IN THE ACCUMULATED LITTER

In general, the FOMM had higher nutrient levels (Table 5) than the Pinus eliottii Engelm plantation, both in terms of macronutrients and micronutrients.

Table 5 - Average content and content of macronutrients and organic carbon in accumulated litter and C/N ratio in Pinus elliotti and Montane Mixed Ombrophylous Forest

Carbon/nutrient	Content ($g.kg^{-1}$)	Content ($kg.ha^{-1}$)	C/N ratio
	Pinus elliotti **Engelm**		
N	7,32	83,39	62,72
P	0,50	6,44	
K	1,05	13,53	
Ca	14,36	185,10	
Mg	2,65	34,15	
C	459,13	5970,00	
	Montane Mixed Ombrophilous Forest		
N	16,48	185,36	25,65
P	1,08	13,61	
K	2,01	25,34	
Ca	17,69	223,07	
Mg	4,96	62,54	
C	422,75	5330,00	

Caldeira (2003) in his study of the Montane Mixed Ombrophilous Forest in the state of Paraná clearly shows that accumulated litter is the main way of transferring nitrogen, potassium and calcium to the soil. Considering only the macronutrients, calcium had the second highest content in the accumulated litter, a result also observed in this study. According to Neves (2000) and Pagano and Durigan (2000), the high variability of K content in leaf litter is closely related to variations in rainfall, which can be explained by its high susceptibility to leaching via leaf and leaf litter washing, due to the fact that potassium does not participate in organic compounds, occurring in soluble form or adsorbed in the cell juice (MARSCHNER, 1997), as the values are lower than those for calcium and magnesium, both in the analysis of the accumulated litter and in the soil analysis, which implies the need to supply other sources, such as the potassium present in the soil solution, which is available in insufficient quantities for the plants.

As far as carbon is concerned, the two forest formations showed some similarity in their results, because regardless of the forest formation, it is well known that accumulated litter benefits the soil.

The carbon and nitrogen contents in Table 2 show a high C/N ratio in the Pinus elliottii Engelm stand, which means intense immobilisation of nitrogen in the protoplasm of the decomposing microorganisms. In practice, tree growth depends on the nitrogen stock in the soil or fertilisation practices when the plantation is still young. Therefore, the desired mineralisation, i.e. the release of mineral nitrogen for absorption by the roots, is not taking place. The fragment of Montane Mixed Ombrophilous Forest, on the other hand, has a C/N ratio within the limits between mineralisation and immobilisation, ranging from 20:1 to 30:1. The above carbon and nitrogen levels show that the Montane Mixed Ombrophilous Forest makes more efficient use of nitrogen than the Pinus elliottii Engelm plantation, in terms of the relationship with the carbon in the litter.

In a study by Sousa et al. (2002) in the Barigui River Alluvial Forest in Araucária, PR, the average organic carbon content found was 421.23 g kg^{-1}, while in the present study, in the Pinus elliottii Engelm and FOMM stand areas, the contents were 459.13 g kg^{-1} and 422.75 g $kg(^{-1})$ respectively, very similar to SOUSA's results.

As for the micronutrients (Table 6), copper was three to six times higher in FOMM than in Pinus Elliottii Engelm (20.60 mg kg^{-1} for FOMM and 5.4 mg kg^{-1} for Pinus Elliottii Engelm). For zinc, iron and manganese, the trend of increasing FOMM contents continued, with values two to three times higher. Therefore, FOMM litter showed higher levels and contents of nutrients absorbed from the soil.

In a study by Caldeira et al. (2008) in the Ombrophilous Dense Forest in Blumenau, the litter accumulated in three stages showed the following decreasing order of micronutrients: Fe > Mn > Zn > B > Cu. This sequence was also observed in another study by the same author (CALDEIRA, 2003) in a Montane Mixed Ombrophilous Forest - PR. In the present study, the sequence observed for both the Pinus elliottii Engelm and FOMM stands was: Zn>Fe>Mn>Cu, and only the zinc levels did not meet this expectation, being much higher than the others. Boron was not analysed.

Table 6 - Average content and content of micronutrients in accumulated litter.

Nutrient	Content (mg kg-1)	Content (kg ha-1)
	Pinus Elliotti **Engelm**	
Ass	5,4	0,06
Zn	1710,00	22,04
Fe	823,00	10,60
Mn	29,60	0,38
	Montane Mixed Ombrophilous Forest	
Ass	20,60	0,25
Zn	2917,00	36,78
Fe	864,00	10,89
Mn	45,40	0,57

In the experiment by Koehler et al. (1987), on a 15-year-old Araucaria angustifolia (Bert.) O. Kuntze plantation in the municipality of Lapa - PR, the average levels of nitrogen, phosphorus, potassium, calcium and magnesium in the leaf litter were 4.0; 0.2; 0.6; 5.2 and 0.8 Kg ha $^{-1}$ year^{-1}. The values found in this study (Pinus elliottii Engelm) for these nutrients were higher than those reported by Koehler (1987).

5.6 QUANTIFICATION OF ORGANIC CARBON IN THE SOIL

Table 7 shows the values for Soil Density (Ds) and CO at the three soil depths in the FOMM and in the Pinus elliottii Engelm stand. The carbon contents decreased along the profile, but with less emphasis in the FOMM, although the highest concentrations were seen in the 0-5 cm depth. When comparing CO between FOMM and Pinus elliottii Engelm stands, there was a higher concentration of carbon in FOMM.

Table 7 - Values of Ds, CO, and total % of CO fixed in the FOMM and in the *Pinus Elliottii* Engelm stand

Depth cm	Ds g/cm^3	Organic carbon g/dm3	Organic carbon Mg.ha^{-1}	% of Total carbon fixed
		FOMM		
0 - 5 cm	0,53	31,16	8,25	17,63
5 - 15 cm	0,59	27,26	16,08	34,37
15 - 30 cm	0,61	24,54	22,45	47,99
Total	-	-	46,78	100,00
		Pinus Elliottii **Engelm**		
0 - 5 cm	0,95	20,64	9,80	20,83
5 - 15 cm	0,95	16,75	15,91	33,82
15 - 30 cm	0,89	15,97	21,32	45,33
Total	-	-	47,03	100,00

The percentages of total carbon fixed were determined using a simple rule of three between the total CO value and the values obtained at each depth.

In the study by Balbinot *et al.* (2003) on *Pinus taeda* L., at 5 years of age, in Rio Grande do Sul, the organic carbon in the soil, up to a depth of 100 cm, was estimated at 227.8 Mg ha^{-1}. In the forest types in this study, soil organic carbon was estimated at 46.78 Mg ha^{-1} for FOMM and 47.03 Mg ha^{-1} for Pinus. However, the soil depth analysed corresponds to approximately one third of the soil depth in the study by Balbinot *et al.* (2003), which does not allow for a reliable comparison of the results. The soil under the FOMM had a low density, probably as a result of the influence of organic matter, which promotes better soil aggregation and, consequently, greater natural porosity. Therefore, the relative quantities at the three depths decrease proportionally in relation to the planting of *Pinus elliottii* Engelm.

On the other hand, the soil under *Pinus elliottii* Engelm showed a density more compatible with the density variation range of most soils.

Monteiro and Delittii (2007) carried out a study on the implications for carbon accumulation in forestry in Cerrado soils, highlighting the large-scale use of the *Eucalyptus sp. and Pinus sp.* genera. The authors concluded that soil carbon decreases along the depth gradient. The work emphasises that the majority of soil carbon is contained in the upper layer. In the 10-25 cm depth, the average values found varied between 10.96 and 16.23 g dm^{-3} in the cerradão areas, between 11.03 and 13.31 g dm^{-3} in the *Eucalyptus* areas and between 7.68 and 11.92 g dm^{-3} in the *Pinus* areas. In the 35 to 50 cm layer, the values were between 8.94 and 12.24 g dm^{-3} in the cerradão areas, between 8.79 and 11.87 g dm^{-3} in the *Eucalyptus* areas and between 7.18 and 10.11 g dm^{-3} in the Pinus areas. The results of this study revealed values below those found in FOMM and stands of_Pinus *elliottii* Engelm, Irati, PR.

In a study on the determination of organic carbon in stands of *Acacia mearnsii* DE WILD planted in Rio Grande do Sul, Saidelles *et al.* (2009) found soil organic carbon levels at a depth of 0 - 20 cm of 9.1 g dm^{-3} at 4 years of age and 10.0 g $dm^{(-3)}$ at 6 years of age, and a quantity of organic carbon in the soil of 26.1 Mg ha^{-1} at 4 years of age and 27.3 Mg ha^{-1} at 6 years of age, thus showing similar behaviour between the two stands. At a depth of 20 - 40 cm, the levels were 6.8 g dm^{-3} at 4 years of age and also 6.8 g dm^{-3} at 6 years of age. At the same depth (20 - 40 cm) the quantities were 21.2 Mg ha^{-1} at 4 years of age and 19.7 Mg ha^{-1} at 6 years of age.

Comparing the results of Saidelles *et al.* (2009) with those of the present study, both for FOMM and for *Pinus elliottii* Engelm stands, the same tendency was observed for the organic carbon content in the deeper horizons to decrease. As for the amount of organic carbon, the values are very similar in the two studies, 46.78 Mg ha^{-1} for Ombrófila Mista Montana and 47.03 Mg ha^{-1} for *Pinus elliottii*, up to a depth of 30 cm, with the sum of Schumacher's values up to 40 cm being very similar, at around 47.0 Mg ha^{-1}. It should be noted that the accumulation of organic carbon was proportionally lower in *Acacia mearnsii*, due to the greater depth of the soil layer.

5.7 CHEMICAL AND PHYSICAL CHARACTERISATION OF THE SOIL

The chemical analysis of the soil (Table 8) revealed higher potassium levels in FOMM compared to the *Pinus elliottii* Engelm. stand, with values ranging from 0.15 to 0.19 cmolc dm^{-3} for Pinus and 0.29 to 0.49 cmolc dm$^{(-3})$ for FOMM. However, the exchangeable bases (calcium and magnesium) showed the opposite, with higher levels in *Pinus elliottii* Engelm. The levels of exchangeable aluminium were higher in the FOMM, which makes the native forest more acidic. Knowing that an increase in the concentration of exchangeable aluminium (Al^{+3}) and low base saturation contribute to a decrease in pH, an increase in acidity may be occurring in the Mixed Ombrophylous Forest due to organic matter decomposition processes.

The FOMM had higher levels of organic matter. Consequently, the area with native forest cover accumulated more carbon than the area planted with Pinus elliottii Engelm.

The physical analysis of the soil (Table 9) showed a predominance of clay and silt fractions, compared to the sand fractions (fine and coarse). Therefore, the soils studied have good water retention and higher CEC.

Table 8- Chemical analysis of the soil at different depths in the Pinus elliottii Engelm plantation and in the FOMM.

Prof./cm	pH	P	Ca	K	Ca + Mg	Mg	A,	H+AI	In	C	MO	SB	CTC
	CaC12	mg.dm^{3}	cmol$_c$.dm^{-3}							gkg^{1}		cmolc.dm^{-3}	
FOMM (0-5)	3,88	4,90	3,16	0,49	3,96	0,80	5,58	19,50	0,21	31,16	53,60	4,66	23,95
FOMM (5-15)	3,87	3,37	1,73	0,33	2,72	0,99	7,34	16,45	0,16	27,26	46,90	3,21	19,50
FOMM (15-30)	3,90	2,51	1,08	0,29	2,27	1,19	7,95	17,47	0,18	24,54	42,21	2,74	20,03
PINUS (0-5)	4,14	2,91	3,31	0,18	4,50	1,19	3,95	13,46	0,20	20,64	35,51	4,88	18,14
PINE (5-15)	4,03	3,17	2,77	0,15	3,46	0,69	4,83	13,86	0,15	16,75	28,81	3,76	17,47
PINE (15-30)	3,99	1,58	2,72	0,19	3,26	0,54	5,34	14,61	0,17	15,97	27,47	3,62	18,06

Table 9- Physical analysis of the soil at different depths in the Pinus elliottii Engelm plantation and in the FOMM.

Prof./cm	Coarse sand	Fine sand	Clay	Silt
		g.kg-1		
FOMM (0-5)	4,75	2,11	60,00	33,14
FOMM (5-15)	1,68	1,01	62,60	34,71
FOMM (15-30)	1,50	0,80	65,00	32,70
PINUS (0-5)	4,37	2,30	56,00	37,33
PINUS (5-15)	4,70	2,50	55,60	37,20
PINUS (15-30)	2,77	1,55	59,20	36,48

The organic matter content was 49.4 g dm$^{(-3)}$ ^{in} Pinus elliottii Engelm and 53.6 g dm^{-3} in FOMM. It can be inferred that, although the organic matter content is similar in Pinus elliottii Engelm and FOMM, the higher nutrient contents show greater soil fertility in the Mixed Ombrophilous Forest area. On the other hand, the exchangeable aluminium content is much higher in the FOMM, due to its greater acidity. It is also important to consider that nutrient levels decrease or stabilise with depth in the Mixed Ombrophilous Forest site.

In a soil survey carried out by the author (OLIVEIRA, 1990 - unpublished) in the same area as this research, in 1990, in a toposequence, to check soil fertility, acidity levels were found to be compatible with those found in this research.

In a study carried out by Scheer (2008) in an area of Alluvial Ombrophilous Dense Forest, in regeneration, in the municipality of Guaraqueçaba, PR, with a predominance of Fluvial Neosols (typical and gleic) and Hystic Cambisols, where values of 0.1 cmolc dm $^{-3}$ were found for K^+ in contrast to 0.49 cmolc dm $^{(-3)}$) in the present study; 1.3 cmolc dm $^{-3}$ for Ca^{+2} compared to 3.16 cmolc dm $^{(-3)}$) in this study. However, the Mg^{+2} content was slightly higher in the aforementioned forest formation (0.9 cmolc dm $^{(-3)}$)) and 0.8 cmolc dm $^{(-3)}$) in the present study. The Al^{+3} content was only 1.9 cmolc dm $^{-3}$ compared to 5.58 cmolc dm $^{-3}$ in the FOMM in the present study, which can be explained by the high acidity of the soil classified as Cambissolo. The phosphorus content was 0.4 mg dm $^{-3}$ compared to 4.90 mg dm $^{-3}$ in this study.

According to a study by Lopes (1983), in soils under conifers, after 20 years of age, there tends to be an increase in the concentration of nutrients, acid pH and lower base removal. There were two findings of a decrease in pH, under Pinus elliottii Engelm at 32 years of age, replacing Eucalyptus sp., and in a Pinus elliottii Engelm plantation, replacing native forest, in Paraná and Santa Catarina, with an increase in exchangeable aluminium levels. The Pinus elliottii Engelm plantation, also 32 years old, has an acidic pH and low natural fertility.

The research by Santos Filho and Rocha (1991) highlights the greater growth of Araucaria angustifolia related to the greater thickness of the solum and the A horizon, in the municipality of Lapa, PR. De Hoogh and Dietrich (1978) also confirm the importance of soil depth in the growth of this species, which is characteristic of the forest formation known as Mixed Ombrophilous Forest. In the present study, the soil classes have in common a poorly developed B horizon, which hinders the growth of the Araucaria specimens, which do not follow the best increments.

These results obtained in the case studies mentioned above reinforce the hypothesis that areas with native forests contribute a greater amount of nutrients to the soil, helping to conserve it.

CHAPTER 6

CONCLUSIONS AND FINAL CONSIDERATIONS

The Montane Mixed Ombrophilous Forest revealed a high potential for accumulating litter and fixing carbon when compared to other tropical forests, and the same finding was observed in the Pinus elliottii Engelm plantation, recommending periodic monitoring of the biomass and carbon stocks accumulated in these forest formations.

It was also found that the Pinus elliotti Engelm plantation did not impoverish the soil, i.e. reduce its fertility. On the contrary, good conditions were observed for the transfer of nutrients from the litter.

The accumulated litter from the FOMM had the highest nutrient (macro and micronutrient) and CO content when compared to the Pinus elliottii plantation, but the highest stock of accumulated litter was observed in the Pinus elliottii Engelm stand.

There was no increase in acidity in the Pinus elliottii Engelm plantation compared to the results in 1990.

Regardless of the forest type in this study, the percentage of total CO fixed in the soil decreases with soil depth.

With regard to the macro and micronutrient levels in the accumulated litter, there were significant differences between the typologies studied. The most significant statistical differences were observed in the N, K, Ca, Mg and Mn contents.

In general, the FOMM soil showed better results in terms of chemical properties, both nutrients and organic matter and organic carbon.

It is hoped that this research can serve as a basis for future work on the importance of the Montane Mixed Ombrophilous Forest and Pinus elliottii Engelm plantations, due to the capacity of these forest formations to accumulate carbon and nutrients, It can also serve as a basis for Clean Development Mechanism (CDM) projects and as an incentive for new investments in Pinus elliottii Engelm forest plantations, with a view to strengthening forest management in terms of improving water quality and fertility and reducing atmospheric air pollution.

CHAPTER 7

BIBLIOGRAPHICAL REFERENCES

ABICHEQUER, A. D. and BOHNER, H. Efficiency of absorption, translocation and utilisation of phosphorus by wheat varieties. **Revista Brasileira de Ciência do Solo**, Campinas, v. 22, n.1, p. 21-26, 1998

ANDERSON, J. M. and SWIFT, M. J. Decomposition in tropical forests. In: SUTTON, S. L.; WHITMORE, T. C.; CNADWICK, A. C. (ed) **Tropical rain forest: Ecology and Management**. Blackwell Scientific Publications, Oxford, p. 287-309, 1983.

ANDRAE, F. H. Effects of nutrient accumulation by aspen spruce and pine on soil properties. **Soil Sci. Soc. Am.** J., v. 46, p. 853-861, 1982.

ANDRAE, F. and KRAPFENBAUER, A. Distribution of fine roots in maritime pine (Podocarpus lambertii) and Brazilian pine (Araucaria angustifolia). In: **PESQUISAS AUSTRO-BRASILEIRAS** 1973, 1982, 1983. Santa Maria: UFSM. p. 56-67, 1983.

ANDRADE, H. and SOUZA, J. J de. **Soils: origin, components and organisation**. Lavras, Minas Gerais: UFLA/FAEPE, 1995.

BALBINOT, R. et al. Inventory of organic carbon in a Pinus taeda plantation at 5 years of age in Rio Grande do Sul, **Revista Ciências Exatas e Naturais**, Vol. 5, no 1, Jan/Jun 2003.

BALBINOT, R. **Carbon, Nitrogen and Isotopic Ratios ô^{13} C and ô^{15} N in Soil and Vegetation of Successional Stages of Submontane Dense Ombrophilous Forest**. Thesis [PhD in Agricultural Sciences] Postgraduate Programme in Forest Engineering, Curitiba: UFPR, 2009.

BARROS, N.F. e NOVAIS, R.F. Relação solo-eucalipto. In: **Folha de Viçosa**, Viçosa: UFV, 1990, 330p. p.14-15, 19-20.

BRAY, J. R. and GORHAM, E. (1964). Litter production in the forests of the world. **Advances in Ecological Research**, New York, V. 2, p. 101-157, 1964.

BRITEZ, R.M.; SILVA, S.M.; SOUZA, W.S. de; MOTTA, J.T.W. Floristic survey in mixed ombrophilous forest, São Mateus do Sul, Paraná, Brazil. **Arq.Biol.Tecnol.**, v.38, n. 4, p. 1147-1161, 1995.

BRUN, E.J. et al. Decomposition of litter produced in three successional stages of deciduous forest in RS. In: SYMPOSIUM OF POST-GRADUATION IN FOREST ENGINEERING, 1, 2001. **Proceedings...** Santa Maria: UFSM, 2001. 1 CD ROM.

BURGUER; D. M and DELLITTI, W. B. C. Biomassa da Mata Ciliar do Rio Mogi-Guaçu. **Revista Brasileira de Botânica,** v. 22, n. 3, 429-435, 1999.

CALDEIRA, M.V.W. **Quantification of biomass and nutrient content in different provenances of black wattle** (Acacia mearnsii **De Wild.)**. Santa Maria, RS.

1998. *96f.* Dissertation (Master's Degree in Forestry Engineering) - Postgraduate Programme in Forestry Engineering, Federal University of Santa Maria.

CALDEIRA, M. V. WATZLAWICK, L. F.; SCHUMACHER, M. V.; BALBINOT, R.; SANQUETTA, C. R. Organic carbon in forest soils. In: **Forests and Carbon**. Chap. 10. Curitiba, 2002. p. 191-214.

CALDEIRA. M. V. W. **Determination of biomass and nutrients in a Montane Mixed Ombrophilous Forest in General Carneiro** - Paraná. Curitiba, 2003. 176f. Thesis [Doctorate in Forestry Sciences] - Agricultural Sciences Sector, Federal University of Paraná.

CALDEIRA, M. V.; MARQUES, R.; SOARES, R. V.; BALBINOT, R. Quantification of litter and nutrients in the Montane Mixed Ombrophilous Forest - PR. In: **Rev. Acad.** V. 5, n.2, p. 101-116, 2007.

CALDEIRA, M. V. V.W. et al. Quantification of litter and nutrients in a dense ombrophilous forest. In: **Semina**:Ciências Agrárias, Londrina, v. 29, n. 1, p. 53-68, Jan/March 2008.

CAMPOS, M.A.A. **Balance of biomass and nutrients in stands of** Ilex paraguariensis **Evaluation in the harvest and in the off-season.** Curitiba, 1991. 106f. Dissertation (Master's Degree in Forestry Sciences) - Agrarian Sciences Sector, Federal University of Paraná.

CASTELA, P. R. (Coord.) **Subproject Conservation of the Araucaria Forest Biome**: diagnosis of forest remnants: final report. Curitiba: FUPEF, 2001. 2 v. il. Project for the Conservation and Sustainable Use of Brazilian Biological Diversity - PROBIO.

CLEVELARIO JÚNIOR, J. **Distribution of carbon and mineral elements in a low-montane humid tropical forest ecosystem.** Viçosa, MG. 1996. 135f. Thesis (Doctorate in Soils and Plant Nutrition) - Federal University of Viçosa.

CUNHA, G.M. et al.. **Dendrometric Characteristics, Nutrients and Carbon in Pioneer and Secondary Species of a Montane Forest Fragment in the Desengano State Park Region.** North of Rio de Janeiro. May 1999 to June 2001.

CUNHA, G.C. **Aspects of nutrient cycling in different successional stages of a Seasonal Forest in Rio Grande do Sul**. Piracicaba, 1997. 86f. Dissertation (Master's Degree in Forestry Sciences) - Escola Superior de Agricultura "Luiz de Queiroz", Universidade de São Paulo.

DECHEN, A. R. and NACHTIGALL, G. R. Micronutrients. In: FERNANDES, M. S. (ed). **Mineral Nutrition of Plants**. Viçosa: Brazilian Society of Soil Sciences, 2006. P. 327-374.

DE HOOGH, R. de and DIETRICH, A. B. **Site evaluation for** Araucaria angustifolia (Bert) O. Ktze in artificial stands. Brasil Florestal, Brasília, v. 10, n. 37, p. 120 1992, 1979.

DE HOOG. R.J. **Site-Nutrition-Growth relantionships of Araucaria angustifolia (Bert.) O. Kuntze in Southern Brazil**. Thesis (Doctorate in Forest Science) - Albert-Ludwigs-Universitat. 1981.

DELITTI, W. B. C. **Comparative Aspects of Mineral Nutrient Cycling in Riparian Forest, Cerrado Field and Planted** Pinus elliottii Engelm. Var. elliottii **(Mogi-Guaçu, SP).** 1994. Thesis (Doctorate in Sciences) - Institute of Biosciences, University of São Paulo, São Paulo.

EMBRAPA. Brazilian Agricultural Research Corporation. **Manual of Soil Analysis Methods**. 2 ed. Rio de Janeiro, National Soil Research Centre, 1997. 212 p.

EMBRAPA. Brazilian Agricultural Research Corporation. **Soil Classification Manual**. Rio de Janeiro, 1999. 412 p.

EMBRAPA. Brazilian Agricultural Research Corporation. **Brazilian Soil Classification System** - Updated. Rio de Janeiro, 1999.

EMBRAPA. Brazilian Agricultural Research Corporation. **Methodology for Estimating Carbon Stock in Different Land Use Systems**. Colombo - PR, 2002.

EMBRAPA. Brazilian Agricultural Research Corporation. **Soil Classification Manual** - Updated. Rio de Janeiro, 2006.

FAQUIN, V. **Nutrição Mineral de Plantas**. Lavras-MG: UFLA, (1994).

FERREIRA, M.M. **Física do Solo**. LAVRAS-MG:UFLA, (1994), pg. 17 - 19.

FIGUEIREDO FILHO. A. Evaluation of diameter increment with the use of dendrometric belts in some species of a Mixed Ombrophilous Forest located in the south of the state of Paraná. In: **Revista Ciências Exatas e Naturais**, Vol. 5, no 1, Jan/Jun 2003

FLORENCE, R. O. and LAMB, D. Influence of stand and site on radiate pine litter in south Australia. **New Zeland Journal of Forestry Science**, Rotorua, 4(3): 502-10, 1974.

FREITAS, A. R. de. Pinus cultivation in Brazil [Preface]. In: KRONKA, F. J. N; et al. **A cultura do Pinus no Brasil**. São Paulo: SBS, 2005.

GALVÃO, F.; et al. (1999) Evaluation of litter deposition in different seral stages of Mixed Ombrophilous Forest. **Anais** V Congresso e Exposição Internacional sobre Floresta. Curitiba. Bio 1068, CD ROOM.

GARAY, I; KINDEL, A; CARNEIRO, R; FRANCO, A. A; BARROS, E; ABBADIE, L.Comparison of Organic Matter and other Soil Attributes between Acacia mangium and Eucalyptus grandis Plantations. In: **Revista Brasileira de Ciência do Solo**. N. 27, p 705-712, 2003.

GARAY, I.; ANDRADE, F. N.; KINDEL, A. Evolution of litter and soil fertility in the Atlantic Tableland Forest: from plantations to native forest. In: CONGRESS

DE ECOLOGIA: ambiente e sociedade, 5., 2001. Porto Alegre. **Abstracts...** Porto Alegre: UFRGS/Ecology Centre, 2001. p. 242.

GARDNER, R. H. and MANKIN, J. B. Analysis of biomass allocation in forest ecosystems of the IBP. In: REICHLE, P. D. **Dynamic properties of forest ecosystems**. Cambridge: Cambridge University Press, 1981. p. 451-497.

GOLLEY, F. B. Forest leaf production. In: MONTGOMERY, G.G. (Ed.). **The Ecology of Arboreae Folivores.** Washington: Smithsonian Institution, 1978. p.17-22.

GONÇALVES, M. A. M. et al. Litter Production in an Atlantic Forest Fragment in the South of the State of Espírito Santo. In: **Anais XII Encontro Latino Americano de Iniciação Científica e VIII Encontro Latino Americano de Pós-Graduação** - Universidade do Vale do Paraíba 2007.

HAAG, H.P. and ROCHA FILHO, J.V.C. & OLIVEIRA, G.D. de. Nutrient cycling in planted Eucalyptus and Pinus forests: the contribution of nutrient species in the blanket. **O solo**, Piracicaba, **70**(2): 28-31, 1978.

HARLOW, W. M.; et al. **Textbook of dendrology**. 6th. New York, 1979.

HOOGH, R. J. de. **Site nutrition growth relationship of Araucaria angustifolia O. Kuntze, in southern Brazil.** 170 f. 1981. Thesis (Doctorate) - Freiburg: Forstwissenschflichen Fakultat.

HUECK, K. **The forests of South America.** São Paulo: Polígloto, 1972.

BRAZILIAN INSTITUTE OF GEOGRAPHY AND STATISTICS - IBGE. **Classification of Brazilian vegetation adapted to a universal system.** Rio de Janeiro, 1991. 124p.

BRAZILIAN INSTITUTE OF GEOGRAPHY AND STATISTICS - IBGE - **Technical Manual of Brazilian Vegetation.** Technical Manuals in Geosciences, n. 1 IBGE, 1992.

BRAZILIAN INSTITUTE OF GEOGRAPHY AND STATISTICS - IBGE. **Brazilian Mesoregions,** 2008.

BRAZILIAN INSTITUTE OF GEOGRAPHY AND STATISTICS - IBGE. **Demographic Census** - general data, 2007.

IRATI, City Hall - Press Department - **general data**, 2006.

IRATI, Prefeitura Municipal de - Departamento de Imprensa- **general data**, 2006. Andrade and Souza (1995, p. 1)

KLEIN, R.M. **The dynamic aspect of the Brazilian pine**. Sellowia, Itajaí, v. 12, n. 12, p. 17-48, 1960.

KLEINPAUL, I. S. et al. Sampling sufficiency for litterfall collection in Pinus elliottii Engelm, Eucalyptus sp. and deciduous forest. In: **Rev. Árvore** [online]. 2005, vol.29, n.6, pp. 965-972.

KOEHLER, et al. Deposition of organic residues (litter) and nutrients in Araucaria augustifolia plantations as a function of site. In: **Revista do Setor de Ciências Agrárias**. Vol 9. Curitiba/UFPR, 1987.

KOZLOWSKI, T.T.; PALLARDY, S.G. **Physiological of woody** plants. 2. ed. San Diego: Academic, 1996. 432p.

KRAPFENBAUER, A. and ANDRAE, F. Inventur einer 17 jaehrigen Araukarienaufforstung in Passo Fundo, Rio Grande do Sul, Brasilien. **Forstwesen**, v. 93, n. 2, p. 70-87, 1976.

KRONKA, F. J. N. et al. **Pinus cultivation in Brazil**. São Paulo: Brazilian Society of Silviculture, 2005.

LACERDA, A.E.B. **Floristic survey and structure of secondary vegetation in a contact area of Dense and Mixed Ombrophilous Forest-PR**. Curitiba, 1999. 114f. Thesis (Master's in Botany) - Biological Sciences Sector, Federal University of Paraná.

LAL, R.; KIMBLE, J.M.; FOLLET, R. F. Pedospheric Processes and the carbon cycle. In: LAL, R.; KIMBLE, J.M.; FOLLET, R. F.; STEWART, B.A (ed) **Soil processes and carbon cycle**. CRC Press LLC, 1998. P. 1-8.

LAMPRECHT, H. **Forestry in the Tropics**. Deutsche:Gesellschaft fur Technische Zusammenarbeit Eschborn 1990.

LARCHER, W. **Plant ecophysiology**. São Carlos: RiMa Artes e Textos, 2000. 531p.

LE BOURLEGAT, J.M.G.; et al. Litter Structure and Mass in Reforestations of Different Ages and Forest Fragments in Northern Paraná. In: **Proceedings of the VIII Congress of Ecology of Brazil, 23 to 28 September 2007,** Caxambu - MG.

LEPSCH, I.F. Influência do cultivo de Eucalyptus e Pinus nas propriedades químicas de solos sob cerrado. **Revista Brasileira de Ciência do Solo**, n. 4, p. 103-107, 1980.

LONGHI, S. J. **A Estrutura de uma Floresta Natural de Araucaria angustifolia (Berth) O. Ktze., no Sul do Brasil.** Curitiba: UFPR, 1980, 198 p. Dissertation (Master's Degree in Forestry Science) - Federal University of Paraná, 1980.

LONGMAN, K. A. and JENIK, J. **Tropical Forest and its environment**, pg. 211-213, Second Edition 1987.

LOPES, M.I.M.S. **Influence of pine cultivation on some characteristics of a dark red latosol originally under cerrado vegetation**. 1983. 90 f. Dissertation (Master's Degree in Agronomy) - Escola Superior de Agricultura "Luiz de Queiroz", Piracicaba, 1983.

MAFRA, Luiz; GUEDES, Sulamita de Fátima Figueiredo; KLAUBERG FILHO, Osmar; SANTOS, Júlio César Pires; ALMEIDA, Jaime Antônio de; DALLA ROSA, Jaqueline. Organic Carbon and Soil Chemical Attributes in Forest Areas. In: **Revista Árvore**, v.32, n. 2, March/April 2008.

MAGALHÃES, L.M.S. and BLUM, W.E.H. Concentration and distribution of nutrients in the leaves of forest species in Western Amazonia. **Floresta e Ambiente**, Seropédica, v. 6, n. 1, p. 127-137, jan./dez., 1999.

MALAVOLTA, E. and VITTI, S. A. **Plant health assessment**; principles and applications, 2ª Ed. Piracicaba: Potafos, 1997.

MALAVOLTA, E. **Manual de nutrição de plantas**. São Paulo: Agronômica Ceres, 2006.

MANASSÉS, J. P. **Inventory of a** Pinus Elliottii **Engelm Plantation**. [unpublished data - field work carried out by the teacher and Forest Engineer of the Colégio Florestal Presidente Costa e Silva] 2007.

MATTOS. P. P.; et al. Dendrochronology of species from the Mixed Ombrophilous Forest in the Municipality of Candói, PR. In: **Pesq. Flor. bras.,** Colombo, n.54, p 153-156, jan./jun. 2007

MARTINELLI, L. A.; et al.. Nitrogem stable isotopic of leaves and soil: tropical versus temperate Forest. In: **Rev.Biogeochimistry. Dordrecht**, v. 46, n. 1/3, p. 45-65, Jul. 1999.

MARSCHNER, H. **Mineral nutrition of higher plants**. 2 ed. San Diego: Academic, 1997.

MEDEIROS, R. P. **Floristic composition of the fragment of Montane Mixed Ombrophilous Forest in the area of the Forestry College** [field work carried out by the teacher and Forestry Engineer of the President Costa e Silva Forestry College 2000 - 2008].

MEDRI, P. S. et al. Comparative analysis of litter accumulation and canopy opening between native forest and commercial reforestation of Pinus taeda L. and Eucalyptus grandis W. Hill Ex Maiden. In: **Anais** IX Congresso de Ecologia do Brasil, 13 to 17 September 2009, São Lourenço - MG.

MIYAZAWA, M. et al. Chemical analyses of plant tissue. In: SILVA, F.C. (Ed). **Manual of chemical analyses of soils, plants and fertilisers**. Brasília: Embrapa Slos, 1999. (Communication for Technology Transfer) p. 171-223.

MONTEIRO, L. L. and DELITTI, W. Silviculture in Cerrado Soils: implications for carbon accumulation. In: **Proceedings of the Brazilian Ecology Congress**, 23 to 28 September 2007, Caxambu - MG.

MORELLATO, L. P. C. Nutrient cycling in two south-east Brazilian forests. I Litterfall and litter standing crop. **Journal of Tropical Ecology**, Cambridge, v.8, n.1, p. 205-215, 1992.

MULLER DOMBOIS, D. and ELLEMBERG, H. **Tentative physiognomic-ecological classification of plants formation of the hearth.** Bericht Uber das Geobot: Inst. Rubel Zurich, v. 37, p. 21-25, 1995-1995.

NEVES, J. C. L. **Biomass production and participation, nutritional and water aspects in clonal eucalyptus plantations in the coastal region of Espírito Santo.** 2000. 191 f. Thesis (Doctorate in Plant Production) - Universidade Estadual do Norte Fluminense, Rio de Janeiro, 2000.

NILSSON, L. O. et *al.* Nutrient uptake and cycling in forest ecosystems - present status and future research directions. **Plant and Soil**, The Hague, v. 168/169, p. 5-13, 1995.

NOVAIS, R.F. and POGGIANI, F. Deposition of leaves and nutrients in pure and mixed Pinus and Liquidambar forest plantations. **Revista IPEF**, Piracicaba, v. 23, p. 57-60, 1983.

O'CONNELL, A.M. and SANKARAN, K.V. Organic matter accretion, decomposition and mineralisation. In: NAMBIAR, E.K.S., BROWN, A.G. (Ed.) **Management of soil, nutrients and water in tropical plantation forests.** Canberra: ACIAR Australia/CSIRO, 1997. p. 443-480. (Monograph; n.43).

OLIVEIRA, L. P. de. **Soil survey in toposequence for the purpose of evaluating natural fertility at the Colégio Florestal Estadual Presidente Costa e Silva in Irati - PR** [unpublished data],1990.

PAGANO, S. N. and DURIGAN, G. Aspects of nutrient cycling in riparian forests of western São Paulo State, Brazil. In: RODRIGUES, R. R. LEITÃO FILHO, H. F. (Ed.). **Riparian forests**: conservation and recovery. São Paulo: EDUSP/FAPESP, 2000. p. 109-123.

PÉLLICO NETTO, S. and BRENA, D. A. **Inventário Florestal**, v. 1. Curitiba, 1977. P. 241.

PRITCHETT, W.L. **Suelos forestales:** propriedades, conservación y majoramiento. Mexico: John Wiley & Sons, Inc., 1990. 634p.

REITZ, R.; KLEIN, R. M.; REIS, A. **Madeiras do Brasil.** Florianópolis. Lunardelli, 1979.p. 320.

REZENDE, MAURO Resende et al. **Pedology and Soil Fertility: Interactions and Applications.** Brasília, DF, 1988.

RICHARDS, P. W. **The tropical rainforest.** London: Cambridge University, 1952. 450p.

ROCHADELLI, R. **The fixation structure of carbon atoms in reforestation (Case study:** Mimosa scabrella **Bentham, bracatinga).** 2001. 86 f. Thesis (Doctorate in Forestry Sciences) - Federal University of Paraná, Curitiba, 2001.

RODRÍGUEZ JIMÉNEZ, L.V.A. **Consideraciones sobre la biomasa, composición química y dinámica del bosque pluvial tropical de colinas bajas. Bajo Calima Buenaventura, Colombia.** Corporación Nacional de Investigación y Fomento Forestal, Bogotá: Serie (Documentación; n.16). 1988. 36p.

SAIDELLES, F. L. F.; et al. Use of equations to estimate organic carbon in Acacia mearnsii De Wild. plantations in Rio Grande do Sul - Brazil. **Rev. Árvore** v. 33, n. 5, Viçosa sept/oct. 2009.

SANQUETTA, C. R. Methods for Determining Forest Biomass. In: SANQUETTA, C.R. et al. **Forests and carbon**. Curitiba: UFPR, p.119-69, 2002.

SANTOS FILHO, A. and ROCHA, H. O. Main soil characteristics that influence the growth of Pinus taeda, in the Second Plantalto Paranaense. **Revista do Setor de Ciências Agrárias**, v. 9, p. 107-111, 1987.

SBS, Brazilian Society of Forestry. Area planted with pine and eucalyptus in Brazil **(Ha) in 2000. São Paulo, 2010. Available at:** http://www.sbs.org.br/area plantada.htm. Accessed on 16/06/2010.

SCHEER, M. B. Decomposition and release of nutrients from leaf litter in a section of Alluvial Ombrophilous Dense Forest in Regeneration, Guaraqueçaba (PR). In: **Floresta**, Curitiba, PR, v. 38, n. 2, April/June 2008.

SCHNITZER, M. and KHAN, S. U. **Soil Organic Matter**. Amsterdam: Elsevier, p. 319, 1978.

SCHLESINGER, W. H. Carbon balance in terrestrial detritus. **Annual Review of Ecology and Systematics**, v.8, p.51-81, 1977.

SCHUMACHER, M. V. Nutrient cycling as the basis for sustained production in forest ecosystems. In: SYMPOSIUM ON NATURAL ECOSYSTEMS IN MERCOSUR: THE FOREST ENVIRONMENT, 1, 1996, Santa Maria. **Proceedings...** Santa Maria: UFSM/ CEPEF, 1996. p. 65-77.

SCHUMACHER, M. V. **Quantification of organic carbon in** Pinus taeda **L. forests of different ages**. Santa Maria: UFSM, 2000 (Research report).

SCHUMACHER, M.V. et al. **Quantification of biomass and nutrient content in the clear-cutting of an Araucaria angustifolia (Bert.) O. Ktze. forest in the Quedas de Iguaçu-PR region**. Santa Maria: UFSM, 2002 (Research report).

SIRTOLI, A. E. et al. **Diagnosis and Recommendations for Soil Management**: theoretical and methodological aspects, Chap. VII, pg. 113-132, UFPR/Sector of Agricultural Sciences 2006.

SPAIN, A.V. Litterfall and the standing crop litter in three tropical Australian rainforests. **The Journal of Ecology**, Oxford, v. 72, p. 947-961, 1984.

SOUSA, S. G . A. de. et al. Understory vegetation of a riparian forest of the Barigui River, Araucaria Paraná. In: National Botanical Congress, 53, Recife, PE. **Abstracts...** Recife: SBB/SRP/UFRPE/UFPE, 2002.

SOUZA, L. C. de (2006) **Nutrient dynamics in precipitation, soil solution and groundwater in three forest typologies on spodosol, on the coast of Paraná.** PhD Thesis, UFPR, Curitiba, PR. 93 p.

STAPE, J. L. Improving the management of planted forests. In: **SIMPÓSIO IPEF...**, 1996, 6., São Pedro. Piracicaba: Instituto de Pesquisa e Estudos Florestais - IPEF, 1996. v. 2, p. 31-37.

STAPE, J. L. IMA (Average Annual Increase) values in m^3/ha/year, age, in the locality of Agudos, SP. In: KRONKA, J.N.; et al. **A Cultura do Pinus no Brasil**. São Paulo: SBS, 2005, 107-108.

STUDENT (pseudonym of William Sealey Gosset, 1876-1937). **The probable error of a mean**. Biometrika, 1908, p.1-25.

SYERS, J. K.; CRASWELL, E. T. Role of soil organic matter in sustainable agricultural systems. In:

LEFROY, R. D. B.; BLAIR, G. J.; CRASWELL, E. T. (Ed.). **Soil organic matter management for sustainable agriculture**. Canberra: ACIAR, 1995. p. 7-14.

TANNER, E. V. J. Litterfall in montane rain forests of Jamaica and its relations to climate. **The Journal of ecology**, Oxford, v. 68, n.3, p. 833-848, 1980.

TEDESCO, M. J. **Analyses of soil, plants and other materials.** 2ª Ed. UFRGS: Porto Alegre, 1995.

TOSIN, J. C. **Influence of Pinus elliottii Engelm, Araucaria angustifólia (Bert) O. KTZE and native forest on the activity of soil microflora.** SCA/UFPR (Master's Thesis) p.111, 1977.

VALE, F. R. do; et al. **Soil Fertility:** dynamics and availability of nutrients. Lavras, Minas Gerais: UFLA/ESAL/ FAEPE, 1994.

VALÉRIO, A. F.; et al. Floristic and Structural Analysis of the Arboreal Component of a Mixed Ombrophilous Forest Fragment in Clevelândia, Southwest Paraná. In: **Rev. Acad. Agrár. Ambient.**, Curitiba, v. 6, n. 2, p. 239-248, Apr./Jun. 2008

VELOSO, H. P.; RANGEL-FILHO, A. L. R.; LIMA, I. C. A. **Classification of Brazilian vegetation, adapted to a universal system**. Rio de Janeiro: IBGE, 1991.

WATZLAWICK, L. F. and CALDEIRA, M. V. W. Estimation of Biomass and Organic Carbon in Pinus taeda L. Stands of Different Ages. **Ver. Biomassa e Energia**, v. 1, n.4, p. 371-380, 2004.

WATZLAWICK, L.F; et al, Inventory of organic carbon in a Pinus taeda plantation, at 5 years of age, in Rio Grande do Sul. In: **Revista Ciências Exatas e Naturais**, v. 5, n. 1, Jan/Jun 2003.

WATZLAWICK, L.F. **Analysing and estimating biomass and carbon in mixed ombrophilous forest and forest plantations using IKONOS II satellite image data**. Curitiba, [2003]. Thesis (Doctorate in Forestry Sciences) - Agricultural Sciences Sector, Federal University of Paraná.

WEBB, D. B. et al. A Guide to species selection for tropical and sub-tropical plantations. In: **CFI Tropical Forestry Papers**.n 15, Oxford, 1980.

WISNIEWSKI, C. (coord.) (1997) **Characterisation of the ecosystem and study of soil-vegetation cover relations in a Pleistocene plain on the coast of Paraná**. Integrated Project - CNPq, Curitiba, 55p.

WISNIEWSKI, C. and REISSMANN, C. B. Deposition of litter and nutrients in Pinus taeda L. plantations in the region of Ponta Grossa - PR. **Arch. Biol. Tecnol.** 39 (2): 435442, June, 1996.

Printed by Books on Demand GmbH, Norderstedt / Germany